Series editor

Lawrie Ryan

Published in 2002 by:
Nelson Thornes Ltd
Delta Place
27 Bath Road
CHELTENHAM
GL53 7TH
United Kingdom

02 03 04 05 / 10 9 8 7 6 5 4 3 2 1

A catalogue record for this book is available from the British Library

ISBN 0 7487 6694 4

Illustrations and page make-up by Wearset Ltd, Boldon, Tyne and Wear

Printed and bound in Spain by Graficas Estella

Introduction

Science for You (Chemistry) has been designed to help students studying Double or Single Science at Foundation level.

The layout of the book is easy to follow, with each new idea on a fresh double page spread. Care has been taken to present information in an interesting way. You will also find plenty of cartoons and rhymes to help you enjoy your work. Each new chemical word is printed in bold type and important points are in yellow boxes.

There are short questions in the text, as well as a few questions at the end of each spread. These help you to check that you understand the work as you go along. The questions at the end of the chapter are there to encourage you to look back through the chapter and apply your new ideas. At the end of each section, you will find lots of past paper questions to help with revision. These are on the coloured pages throughout the book.

At the end of each chapter, you will see a useful summary of the key facts you need to know. You can test yourself by answering Question 1 that follows each summary.

As you read through the book, you will come across these signs:

This shows where there is a chance to use computers to help you find information or view simulations.

This shows where experiments can be done to support your work.

(The instructions are on sheets in the Teacher Support CD ROM.)

There is an extra section at the end of the book. Here you can get help with your Coursework, Revising and doing your exams, and Key Skills.

Using this book should make chemistry easier to understand and bring you success in your exam.

Finally, I hope you'll have fun studying chemistry, after all, most of us enjoy the things we're good at!

Good luck!

Lawrie Ryan

Contents

Section Three

Chemical reactions and their useful products

Section Four

Chemical patterns and bonding

CHAPTER 1

Foundations of Chemistry

1a Particles

Diffusion gives us great smells (sometimes!).

Has your mouth ever watered when you smell good cooking?
Have you ever wondered how the smell reaches you?
The smell arrives at your nose as tiny food particles
– too small to see. The particles are given off from the hot food.

a) Name 3 types of shop that you could recognise by their smell.

Everything is made up from particles – even you!

When the particles of one substance move through and mix with particles of another substance, we call it **diffusion**.

Diffusion gives us evidence that particles exist.

b) List some cosmetics that work by diffusion.

Solids

We believe that the particles in a solid are set out as shown below:

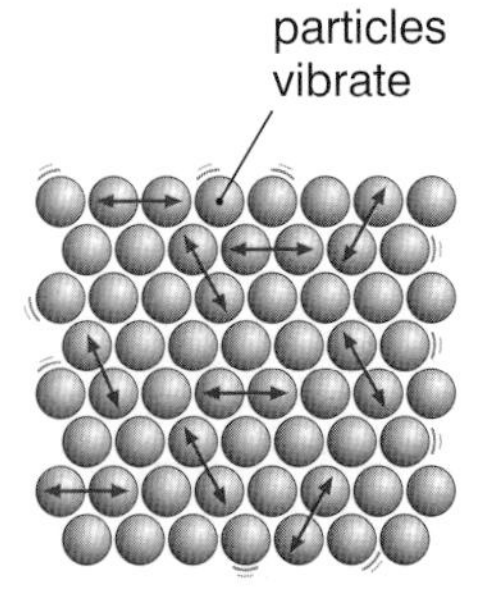

Squashed together, the particles are tight,
Some think they're still – but that's not right.
Even though you can't see them shaking,
Believe it or not they are vibrating.

c) Do the particles in a solid move at all? Explain your answer.

d) Think of one property of solids:
Can you explain this using the 'particle model' shown above?

e) What do you think happens to the particles in a solid when it melts?

Liquids

As we heat a solid, its particles start vibrating more and more quickly.
At its melting point, the particles become free to move around.
Look at the diagram below:

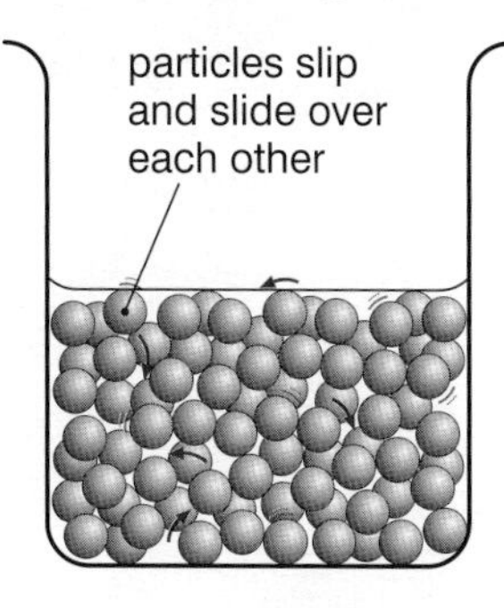

Particles mingling and inter-twined,
There's not much room in here to find!
The forces of attraction are still quite strong,
But particles are free to move slowly along.

f) What do we call it when we cool a liquid down to form a solid?
g) Describe what happens to the particles in question f).

Gases

When you boil a kettle, what comes out of the spout?
What is happening to the water particles?
The particles in a gas have a lot more space in between them.
They zoom around very quickly.
Look at the diagram below:

particles move very quickly in all directions; as the particles bash against the walls of the container, they exert a force that causes pressure

Particles whizzing everywhere,
Zooming around, they just don't care!
Speeding left and speeding right,
With gaps in between they're amazingly light.

h) Write down the names of 3 gases

Remind yourself!

1 Copy and complete:

The particles in a ...... are packed ...... together.
They ...... 'on the spot'.
In a ...... there is a little more space and particles become ...... to slip and slide over each other.
In a gas there is lots of ...... between particles, and they move around very ......

2 Draw a concept map to show all that you know about solids, liquids and gases.
Try to include the words:

particles, melt, boil, condense, freeze

3 Imagine you are a water particle in a kettle.
Describe what happens to you when someone makes a cup of tea.

1b Hydrogen gas

Did you think of hydrogen gas when you answered question h) on the previous page? You have probably come across hydrogen when testing gases in science lessons.

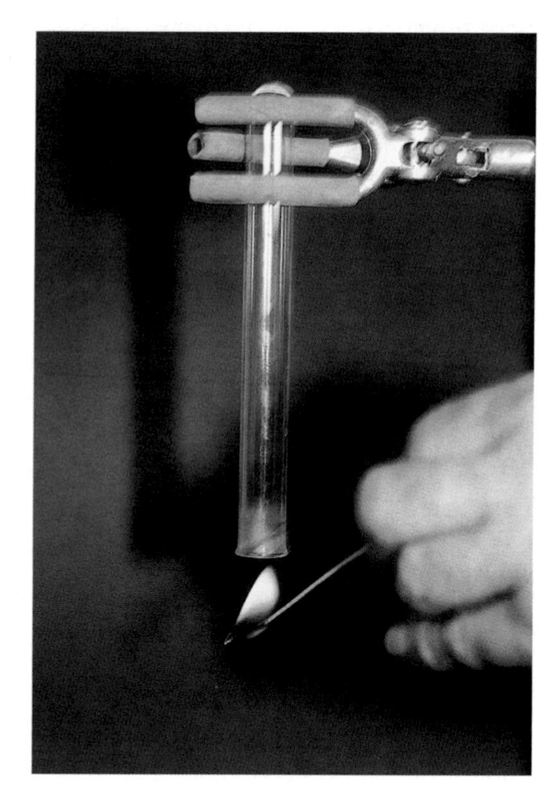

Testing for hydrogen gas.

> **Test for hydrogen gas** – it burns with a squeaky pop when you put a lighted splint near the mouth of a test tube of the gas.

a) What happens when you test hydrogen gas with a lighted splint?

Hydrogen mixed with oxygen from the air forms an explosive mixture. That's why we get the squeaky POP!

Hydrogen is the lightest of all gases.
So it was used to fill the first air-ships.
These were giant canvas balloons. They had engines to drive them through the air and large passenger compartments.
Hydrogen gave the huge structures the lift they needed to fly.

However, it was another property of hydrogen that proved to be fatal. Think of the test for hydrogen:
Hydrogen gas is very flammable.
Look at the photo below:

The Hindenburg disaster in 1937.

b) Think of some reasons why the hydrogen might have exploded.

c) Which gas in the air was the hydrogen reacting with?

d) Which gas are modern air-ships filled with?
(Hint: its name also begins with H.)

Uses of hydrogen

As you have seen, hydrogen gas is very flammable.
So we can use it as a fuel in rockets.
Every time a space shuttle is launched, hydrogen helps to blast it into space.

e) What might be a problem with using a gas as a fuel?

Look at the diagram below:

f) How can you get as much hydrogen as possible into the shuttle's fuel tanks?

Hydrogen gas is made up from 2 hydrogen **atoms** joined together.
It makes a hydrogen **molecule**.

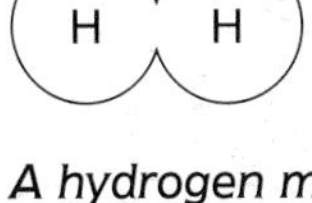

A hydrogen molecule.

Molecules are groups of 2 or more atoms bonded together.

g) Draw 2 boxes.
In one box show the hydrogen molecules when it is a gas.
In the other draw the hydrogen molecules in the space shuttle's fuel tank.

h) Hydrogen is also used in space-craft in fuel cells.
Do some research and write a short report on fuel cells. ICT

Remind yourself!

1 Copy and complete:

We use a ...... splint to test for hydrogen gas. The hydrogen burns with a squeaky ......

Hydrogen is the ...... of all gases.

2 a) What is made when hydrogen reacts with oxygen?

b) Why does your answer to a) make hydrogen a good fuel for the future?

3 Write a short newspaper report on the Hindenburg disaster.

1c Oxygen gas

Have you ever been ill and had to be given oxygen gas? People having an asthma attack can breathe more easily if given more oxygen. Did you know that only about 20% of air is oxygen? So your lungs can absorb more oxygen by breathing gas from the cylinder.

Look at some other uses of oxygen in these photos:

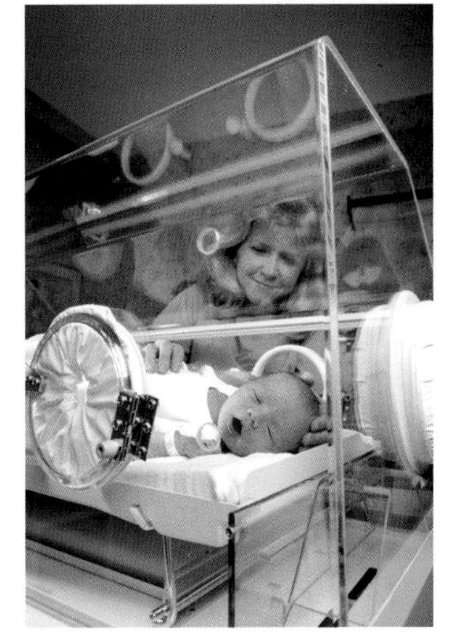

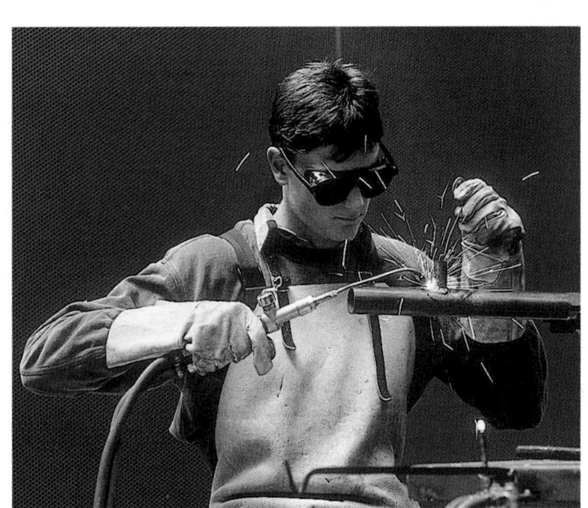

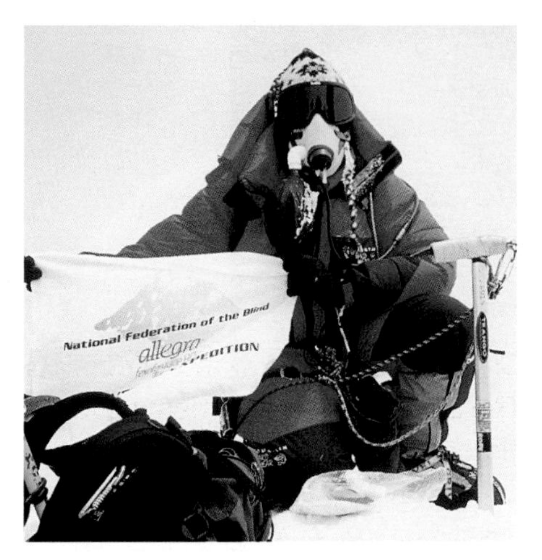

Oxygen is the gas we need when we breathe or when we burn things.

a) What fraction of the air is oxygen (roughly)?
b) Which gas makes up most of the air?
c) Look at the photos above:
Think of two other places where we use extra oxygen gas to help us breathe.

Things burn very well in pure oxygen.
In 1967 this was unlucky for the astronauts in Apollo 1.

They were practising for their voyage to space when faulty electrical equipment made a spark. They were in an atmosphere of pure oxygen inside their space-ship. The 3 men were killed instantly in the flash fire that followed.
They don't use pure oxygen inside space-ships now.

The test for oxygen is based on the fact that things burn well in the gas:

Oxygen gas re-lights a glowing splint.

Like hydrogen, oxygen forms 'two-atom' molecules:
Hydrogen and oxygen are both examples of **chemical elements**.
Look at the diagrams below:

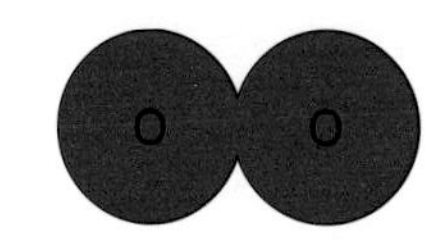

An oxygen molecule.

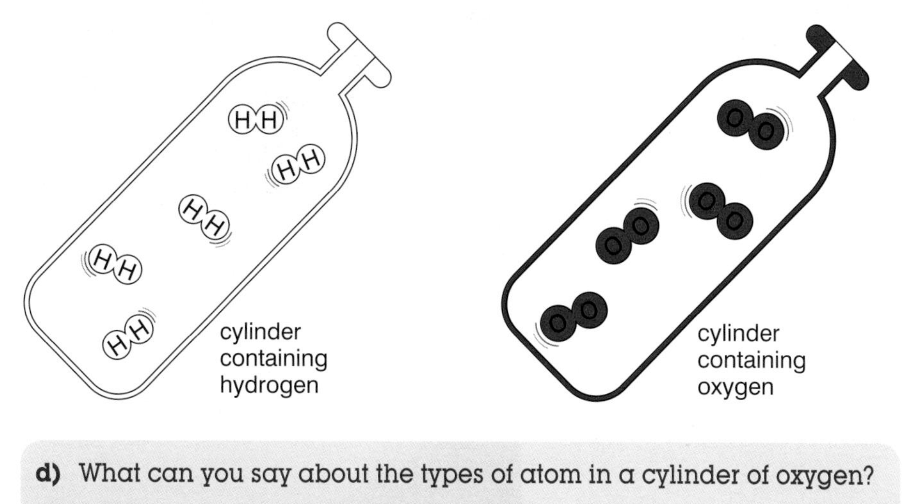

d) What can you say about the types of atom in a cylinder of oxygen?

Elements contain only one type of atom.

Chemists use short-hand to describe atoms and molecules.
The atoms of each element have a symbol.
Here are some common symbols:

Element	Symbol
Hydrogen	H
Oxygen	O
Nitrogen	N
Carbon	C
Helium	He
Chlorine	Cl
Magnesium	Mg
Zinc	Zn
Sodium	Na
Iron	Fe
Lead	Pb
Tin	Sn

e) Choose 10 symbols and test a partner to see if they can remember their symbols.

Remind yourself!

1 Copy and complete:

We use oxygen gas every time we …… in.
It is also needed to …… things.
In the test for oxygen, a …… splint will re-……

2 Look at the table above:
Sort the elements into sets that are based on how their name relates to their chemical symbol.

3 Find out how oxygen is used in making steel.
Make a leaflet on steel making for a Y7 pupil.

1d Carbon dioxide gas

Do you like fizzy drinks? Have you ever had that weird feeling when the bubbles come back down your nose? That's carbon dioxide!

Carbon dioxide gas is dissolved in fizzy drinks under pressure. When you release the pressure, by opening a can or bottle, the gas escapes.

The higher the pressure, the more gas dissolves in water.

If you don't screw the top on your fizzy drink tightly, it goes flat as the carbon dioxide gas escapes.

a) Sketch a graph to show how much carbon dioxide dissolves as we change the pressure:

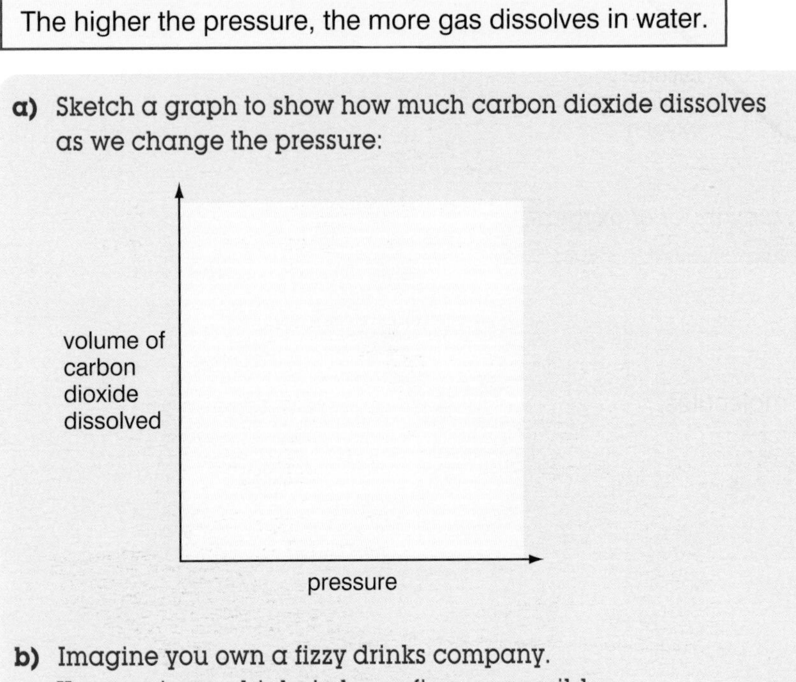

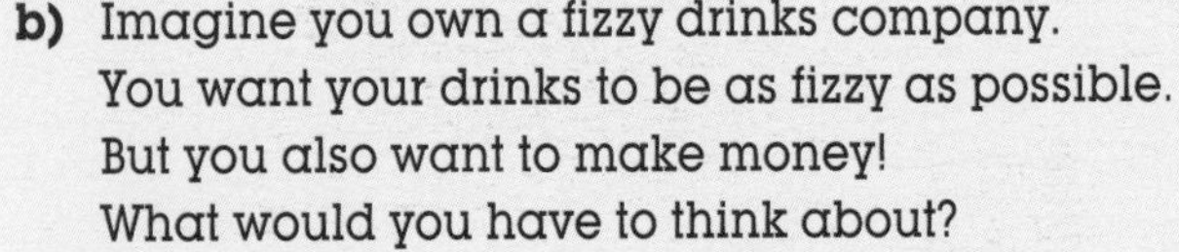

b) Imagine you own a fizzy drinks company.
You want your drinks to be as fizzy as possible.
But you also want to make money!
What would you have to think about?

On the previous page, we saw that all the elements have a symbol.

You might know the **chemical formula** of carbon dioxide. It is **CO_2**.

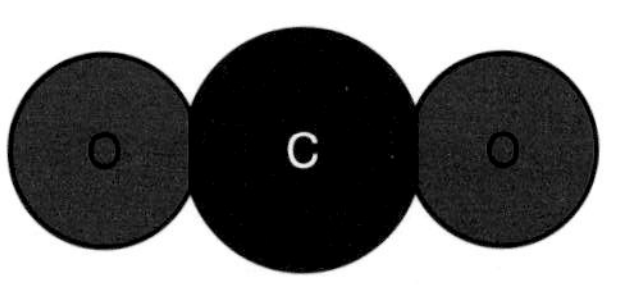

A carbon dioxide molecule.

c) Which two elements make up carbon dioxide?

d) How many atoms are there in a CO_2 molecule?

Carbon dioxide is an example of a **compound**.

Compounds are made up from more than one element.
They contain more than one type of atom bonded together.

e) Give the name and formula of another chemical compound.

Have you ever made carbon dioxide in an experiment?
The test for carbon dioxide uses limewater.

Carbon dioxide gas turns limewater milky / cloudy.

Look at the properties of carbon dioxide below:

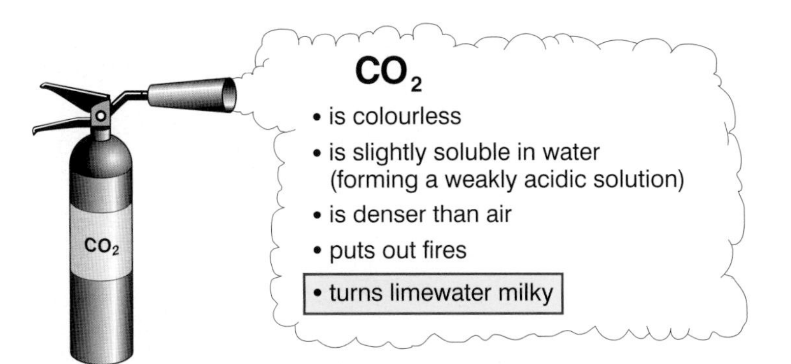

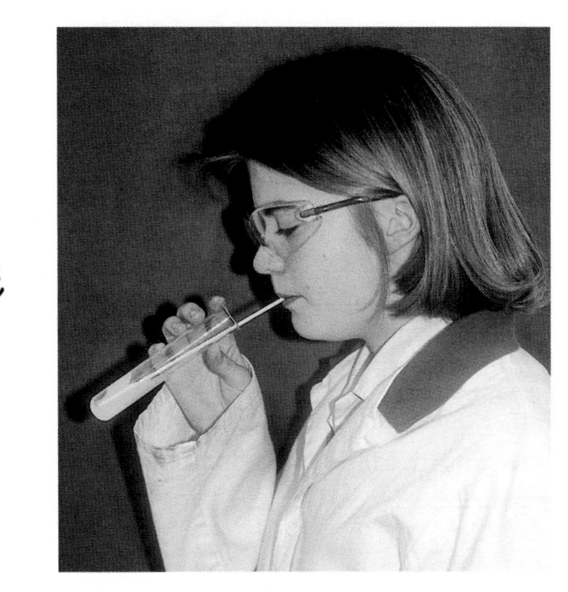

Limewater turns milky/cloudy when we bubble carbon dioxide gas through it.

Look at the pictures that show some uses of carbon dioxide:

f) Describe how carbon dioxide features in each picture above.

g) Why is the foam important at the aeroplane fire?

h) Champagne is a type of wine. Find out how it gets its 'fizz'. Look at the top of the champagne bottle opposite: Explain why the wire is used.

ICT

Remind yourself!

Copy and complete:

1 The chemical formula of carbon dioxide is …… The gas turns …… milky / cloudy.
Things do not …… in carbon dioxide which explains its use in fire ……

2 Look at the list of formulae below:
NH_3 N_2 H_2 CH_4 Cl_2 S_8 H_2S

a) Draw pictures to show the atoms bonded together in each molecule.

b) Draw a table showing which are elements and which are compounds.

1e Chemical reactions

As you read this book, chemical reactions are keeping you alive. Inside cells in your body, sugar (glucose) is reacting with oxygen. The reaction forms carbon dioxide and water.

The substances reacting together are called **reactants**.
The substances formed in the reaction are called **products**.

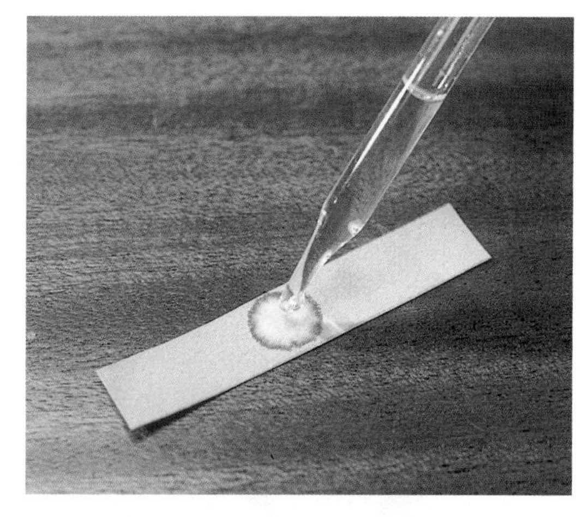

Testing for water with blue cobalt chloride paper.

a) Name the two reactants in your cells.

b) What are the two products in this reaction?

We can show a chemical reaction by a **word equation**.
For the reaction in our cells, the word equation is:

sugar + oxygen → carbon dioxide + water
reactants **products**

We get rid of the products of the reaction when we breathe out.

c) How could you show that water is one of the products when we breathe out?
(Water turns blue cobalt chloride pink or white anhydrous copper sulphate blue).

d) How could you show that carbon dioxide is a product when we breathe out? (Hint: Look on the previous page).

Testing for water with anhydrous copper sulphate.

We saw on page 9 that hydrogen is used as a fuel.
When it burns in air, it reacts with oxygen gas.
Water (hydrogen oxide, H_2O) is made in the reaction.
Combustion is the chemical name for burning.

e) What are the reactants when hydrogen burns?

f) Name the product of this reaction.

g) Write a word equation to show hydrogen burning.

When carbon burns, it forms carbon dioxide.

h) Write a word equation to show carbon burning.

i) What do we call this type of reaction?

Carbon burns in a barbecue.

Balancing equations

You can also show equations using symbols.
In the last reaction, carbon (C) reacted with oxygen (O_2) to give carbon dioxide (CO_2). This is shown as:

$C + O_2 \rightarrow CO_2$

j) Count the number of carbon and oxygen atoms before and after the reaction. What do you notice?

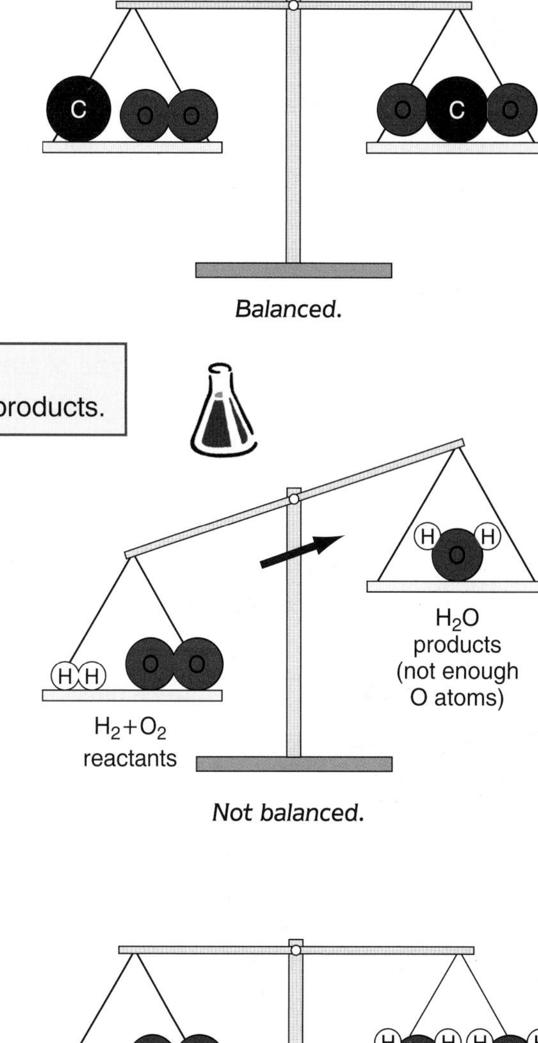

Balanced.

Not balanced.

> No new atoms are made or destroyed in a chemical reaction.
> The total mass of the reactants is the same as the total mass of the products.

Sulphur (S) also burns in air to give sulphur dioxide (SO_2).

k) Write a symbol equation to show the reaction above.

Now think about hydrogen reacting with oxygen:
Both hydrogen and oxygen gases exist as two-atom molecules.
These are called **diatomic molecules.**
When they react:

$H_2 + O_2 \rightarrow H_2O$ ✗

l) What do you notice about the number of H and O atoms on either side of the arrow? Is the equation 'balanced'? Which atom are we short of?

To balance equations we can ***never change a formula***.
We can't just change H_2O to H_2O_2.
(In fact H_2O_2 is hydrogen peroxide – used to bleach hair blonde!)
But we can put numbers in front of formulae.
For example, 2 H_2O means we have 2 molecules of H_2O.
In 2 molecules of H_2O we have 4 H atoms and 2 O atoms.
This means that we have enough O atoms, but now we need 2 more H atoms on the left hand side. We can do this by putting a big 2 in front of the H_2:

$\mathbf{2\ H_2 + O_2 \rightarrow 2\ H_2O}$ ✓

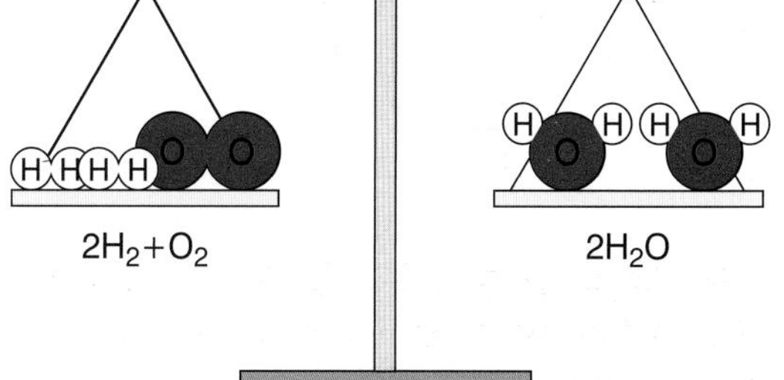

Balanced at last!

m) Count the number of H and O atoms on each side of the equation. Is the equation balanced now?

ICT

Remind yourself!

Copy and complete:

1 Reactions can be shown by a w...... equation. No new atoms are ...... or ...... in a chemical reaction.
If you use formulae the equation must be ......

2 Finish these equations:
a) iron + oxygen → ....
b) magnesium + oxygen → ....

3 Balance these equations:
a) $H_2 + I_2 \rightarrow$ ...... HI
b) ...... Na + $Cl_2 \rightarrow$ ...... NaCl

Summary

All matter is made up from tiny particles, too small to be seen.
In solids the particles are packed together tightly and vibrate.
In liquids the particles are still close together but they are free to slip and slide over each other.
In gases the particles zoom around and have lots of space between them.

Atoms.

Groups of **atoms** bonded together are called **molecules**.
Substances made of only one type of atom are called **elements**.
Substances containing different types of atoms (elements) are called **compounds**.

Molecule of an element.

We can show chemical reactions by a word or symbol equation. These show what we start with (**reactants**) and what is formed (**products**).

Reactants → Products

Molecule of a compound.

Questions

1 Draw 3 boxes to show how the particles are arranged in:

a) a solid b) a liquid c) a gas

2 a) Describe how you can smell some strong perfume from several metres away.

b) What do we call this process?

c) Think of another example that shows this process.

3 Which of these substances are elements and which are compounds?

a) fluorine (F_2)

b) chlorine (Cl_2)

c) water (H_2O)

d) phosphorus (P_4)

e) phosphine (PH_3)

f) sodium chloride (NaCl)

g) calcium carbonate ($CaCO_3$)
Look at the compounds in your answer above:

h) Say how many elements there are in each of the compounds.

i) Say how many atoms there are in the molecules a) to e) above.

4 When hydrogen (H_2) burns in chlorine gas (Cl_2), hydrogen chloride gas (HCl) is made.

a) Which are the reactants and which are the products in the reaction.

b) Write a word equation that shows this reaction.

c) Which has the greater mass, the hydrogen and chlorine, or the hydrogen chloride.

d) Draw pictures of all the molecules involved in the reaction.

e) Write a balanced chemical equation for the reaction.

5 Draw a concept map that links together the following words:

element
compound
atom
molecule

Make sure you label the lines joining the words.

Section One
Metals

**In this section you will find out more about metals and how important they are.
You will see where they are in the Periodic Table, how we extract metals, and about the reactions of acids to make compounds of metals.**

CHAPTER 2

Groups of Metals

2a Properties of metals

Metals are very important in our lives.
Just try to imagine a world without metals.

a) What do you think would be the biggest change in your life if we no longer had any metals?

The strings of this guitar are made from steel.

Here is a reminder of the properties of most metals:

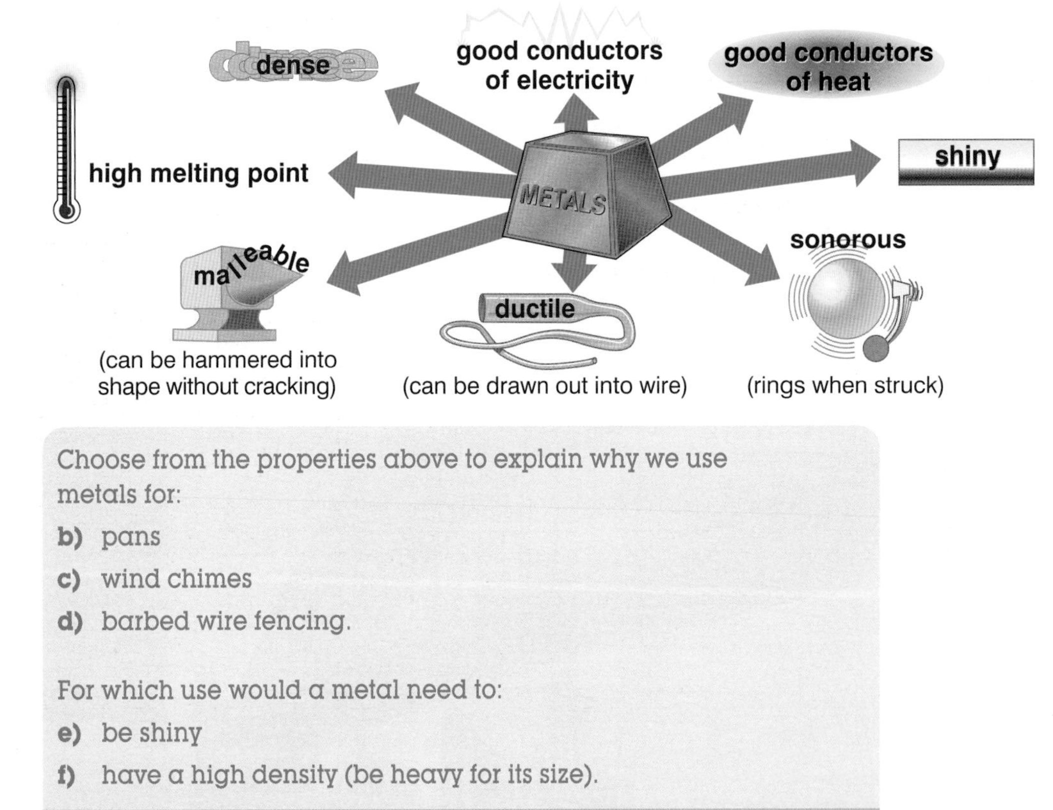

These are the magnetic metals:

iron
cobalt
nickel

Choose from the properties above to explain why we use metals for:

b) pans
c) wind chimes
d) barbed wire fencing.

For which use would a metal need to:

e) be shiny
f) have a high density (be heavy for its size).

In the last chapter we saw how each element is made from one type of atom.
We have 92 different types of atom found naturally.
(And scientists have made about another 20 in the last 60 years.)
But there are millions of different chemical substances on Earth.

g) Explain how we can have so many different substances.

Arranging the elements

Most of the elements are metals.
In fact, ***over three quarters are metals***.

h) What is the missing number (not %) below:
Of the 112 elements, over ...... are metals.

Chemists have always been interested in any patterns they could find to help sort out the elements.
Early attempts, in the mid-1800's, put the atoms in order of mass.
Then similar elements were lined up on top of each other.
This worked well most of the time. But some elements, such as potassium and argon, appeared in the wrong places.
(You can read how this has been solved on page 192.)

This table of elements is called the **Periodic Table**.
Look at the Table below:
It has been divided into metals, non-metals and a few elements that have some properties of both. These are called semi-metals or metalloids.

Dmitri Mendeleev was a chemistry teacher in Russia! He devised the Periodic Table. You can read more about him on page 191.

The Periodic Table (showing the first 86 elements and atomic masses).

- metals
- non-metals
- semi-metals or metalloids

			1 H hydrogen														4 He helium
7 Li lithium	9 Be beryllium											11 B boron	12 C carbon	14 N nitrogen	16 O oxygen	19 F fluorine	20 Ne neon
23 Na sodium	24 Mg magnesium											27 Al aluminium	28 Si silicon	31 P phosphorus	32 S sulphur	35 Cl chlorine	40 Ar argon
39 K potassium	40 Ca calcium	45 Sc scandium	48 Ti titanium	51 V vanadium	52 Cr chromium	55 Mn manganese	56 Fe iron	59 Co cobalt	59 Ni nickel	64 Cu copper	65 Zn zinc	70 Ga gallium	73 Ge germanium	75 As arsenic	79 Se selenium	80 Br bromine	84 Kr krypton
85 Rb rubidium	88 Sr strontium	89 Y yttrium	91 Zr zirconium	93 Nb niobium	96 Mo molybdenum	99 Tc technetium	101 Ru ruthenium	103 Rh rhodium	106 Pd palladium	108 Ag silver	112 Cd cadmium	115 In indium	119 Sn tin	122 Sb antimony	128 Te tellurium	127 I iodine	131 Xe xenon
133 Cs caesium	137 Ba barium	139 La lanthanum	178 Hf hafnium	181 Ta tantalum	184 W tungsten	186 Re rhenium	190 Os osmium	192 Ir iridium	195 Pt platinum	197 Au gold	201 Hg mercury	204 Tl thallium	207 Pb lead	209 Bi bismuth	210 Po polonium	210 At astatine	222 Rn radon

Remind yourself!

1 Copy and complete:

Most of the chemical elements can be split into 2 sets – the metals and the ......
Over ......% of the elements are metals.

2 Look at the Periodic Table above:
Find any examples where the elements are not in order of atomic mass.

3 Find out the uses of one particular metal. Then draw an advert explaining the uses of your metal.

2b Group 1 – The alkali metals

The metals in this first group (column) of the Periodic Table are ***very reactive***.

Look at their symbols and names below:

Li	lithium
Na	sodium
K	potassium
Rb	rubidium
Cs	caesium

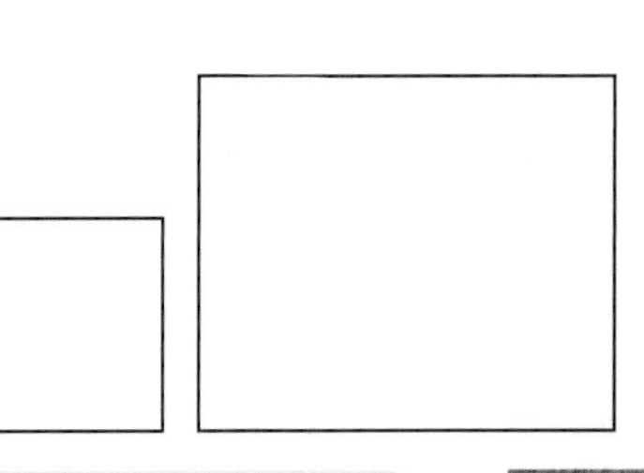

a) Which of the Group 1 metals have symbols that don't look at all like their name in English?
Try to find out where these symbols came from.

When was the last time you came across a Group 1 metal?
The metals themselves are too dangerous for you to use.
But you have probably used their compounds today.
Did you brush your teeth this morning? Next time you do, look on the toothpaste tube. If it contains fluoride, it is usually in a sodium compound.
Or have you eaten food with salt on today?
Table salt is sodium chloride.

Many toothpastes contain sodium fluoride to protect your teeth from decay.

b) Which Group 1 metal compounds are in the tooth-paste shown in the photo?

c) Name another common Group 1 metal compound.

Properties of Group 1 metals

These are a strange group of metals.
They are certainly not typical of the metals we see every day.
Look at the photo opposite:

The group 1 metals:
- are soft (they can be cut with a knife)
- are not dense (lithium, sodium and potassium actually float on water)
- have to be stored under oil.

Sodium is a soft metal.

d) Name a metal that is a 'typical' metal.

e) Why do you think the Group 1 metals are kept in jars of oil?

Reactions of the Group 1 metals

Potassium reacting with water.

Some groups in the Periodic Table have special names as well as their group number.
The Group 1 metals are called the **alkali metals**.
This is because of their reactions with water.
Look at the photo opposite:

They react vigorously, giving off ***hydrogen*** gas.
When you test the solutions left at the end, they are ***alkaline***.
Here is the word equation for sodium:

sodium + water → sodium hydroxide + hydrogen

The alkali metals can be cut with a knife.

f) How would you do the test to see if the solution formed was an alkali?
g) Which compound dissolved in the water to make it alkaline?
h) If you collected the gas, how could you show it was hydrogen?

Once you know one reaction for a group in the Periodic Table, you know it for all the elements in the group.
So for lithium we would have:

lithium + water → lithium hydroxide + hydrogen

i) Write the word equation for potassium reacting with water.

Sodium reacting with oxygen gas.

The alkali metals also react with non-metals.
Look at the photo opposite:
The sodium is reacting with oxygen.
It forms a white solid called sodium oxide.

j) Write a word equation for this reaction.

The alkali metals react with non-metals to form ***white solids***.
These solids ***all dissolve in water***.

Remind yourself!

1 Copy and complete:

Group 1 metals are called the ...... metals.
They have a ...... density and are stored under ...... They are very r......
For example, they react vigorously with water giving off ...... gas and leaving an ...... solution.
They react with non-metals, forming ...... solids that are ...... in water.

2 a) Write a word equation for caesium reacting with water.

b) Draw a poster to show what you think would happen in this reaction.

2c Transition metals

Now these elements really are ***'typical' metals!***
Most have all the usual properties of metals on page 18.

How do you get to school? If it's any way other than walking, you can thank a transition metal for getting you there. Most cars, bikes, buses and trains are made from steel. And steel is almost completely made from **iron (Fe)** – one of the **transition metals**.
We find the transition metals in the central block of Periodic Table:

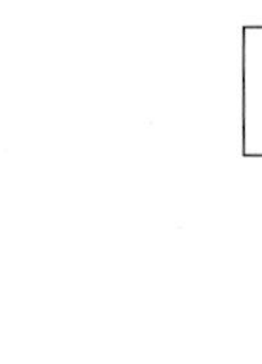

Sc scandium	Ti titanium	V vanadium	Cr chromium	Mn manganese	Fe iron	Co cobalt	Ni nickel	Cu copper	Zn zinc
Y yttrium	Zr zirconium	Nb niobium	Mo molybdenum	Tc technetium	Ru ruthenium	Rh rhodium	Pd palladium	Ag silver	Cd cadmium
La lanthanum	Hf hafnium	Ta tantalum	W tungsten	Re rhenium	Os osmium	Ir iridium	Pt platinum	Au gold	Hg mercury

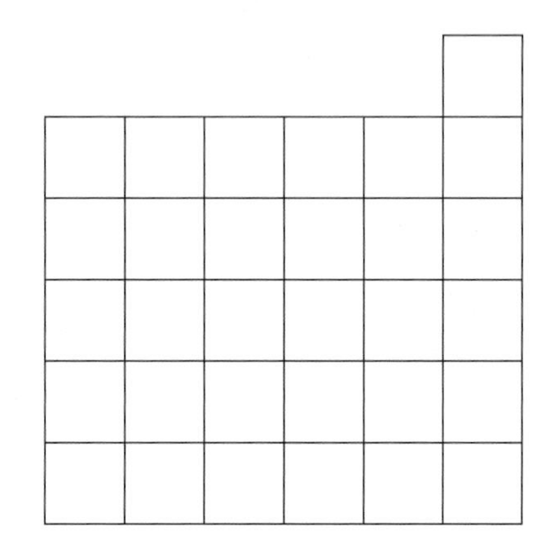

a) Which of the metals shown in the central block have you never heard of? Which ones do you know? Draw a table to show your answers.

b) What are silver, gold and platinum all used for? Explain why.

c) Which metal in the table above does not have all the typical properties of a transition metal? (Hint: It's a liquid at room temperature.)

Transition metals are strong, tough and hard.

The transition metals have many uses.
Have you any coins on you? If you have, you'll be carrying a lot of **copper (Cu)**. Even a so-called 'silver' coin like a 50p piece is three quarters copper (the rest is nickel – not silver!).

d) Think of two other uses of copper.

e) What makes copper a good metal for these uses and for coins?

These are made from copper.

Copper is a ***good conductor of electricity***.
It is used in the wiring at home and in cables.
Have you ever seen a copper pan and kettle? Nowadays most pans are made from stainless steel. Stainless steel is made mainly from iron with a little chromium and nickel added.

f) What properties must a good pan have?

Transition metals:
- are strong, tough and hard,
- are good conductors of heat and electricity,
- have high melting points (apart from mercury!),
- can be hammered into sheets and drawn out into wires.

Look at some uses of transition metals below:

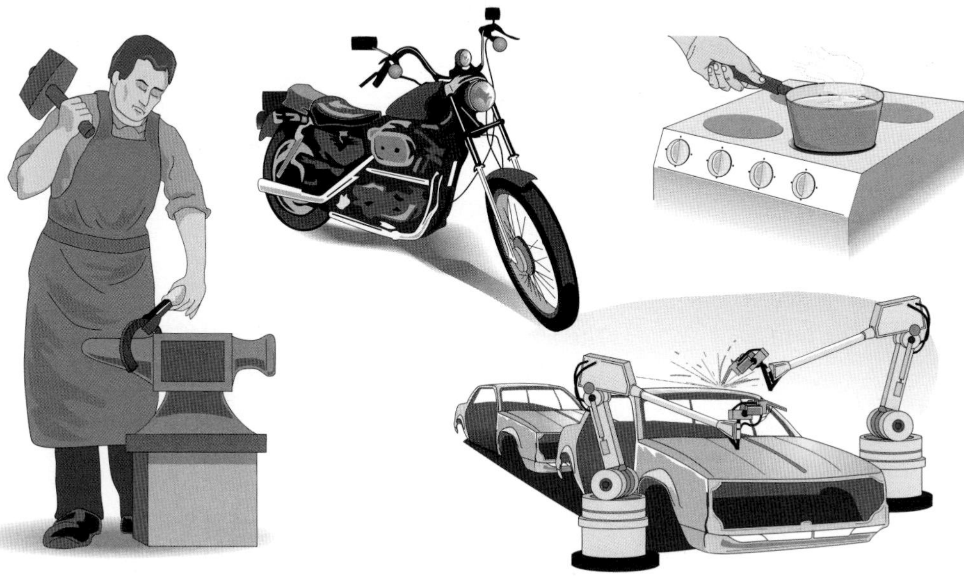

g) Link the uses shown above to the properties at the top of the page.

Transition metals are not very reactive.
This makes them useful too.

The transition metals are much ***less reactive*** than the alkali metals.

They don't react with oxygen or water very quickly.
Unfortunately iron does **rust** over time. Then the rust just crumbles away.
However, we can protect iron from rusting. (See page 37.)

Have you ever seen a green roof like the one opposite?
Copper weathers very slowly. The metal forms a green coating of a copper compound. Unlike the alkali metals:

The transition metals form ***coloured*** compounds.

Pottery glazes also rely on transition metal compounds for their colours.

The copper on this roof takes years to react in the air.

Remind yourself!

1 Copy and complete:

The transition metals are good ...... of heat and ...... They are strong, ... and hard.
They are not very ...... with water or oxygen.
When they do react, they form ...... compounds.
These are useful in staining the ...... on pottery.

2 Make a table with iron and copper in it, showing as many uses of each metal as you can.

3 Do some research to find out about the uses of transition metals as **catalysts**. Show your results in a table.

Summary

Metals make up over three quarters of all the elements.
In the Periodic Table, they are found in Groups 1 and 2, the central block and some in the right-hand block.

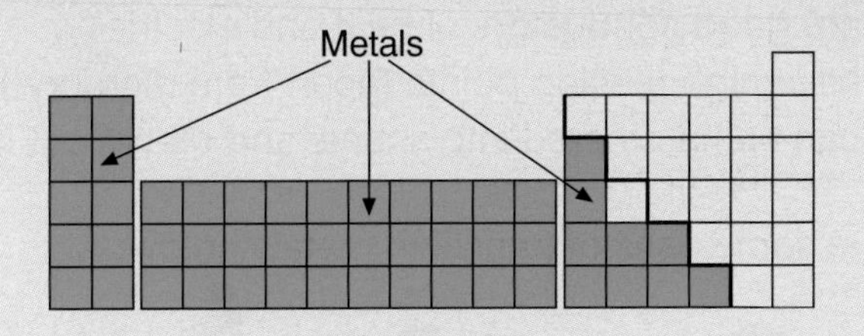

The **Group 1** metals are called the **alkali metals**.
For metals, they have low densities and are soft.
They are very reactive metals.
When dropped on to water they fizz around the surface giving off hydrogen gas. They form an alkaline solution of the metal hydroxide.
Most of the compounds of alkali metals are white.
They all dissolve in water.

The alkali metals. (Sodium vapour is used in yellow street lights)

In the central block of the Periodic Table we find the **transition metals**.
These have the properties of typical metals.
They have:

- high melting points,
- are good conductors of heat,
- are good conductors of electricity,
- can be hammered into shapes (are malleable),
- can be drawn out into wires (are ductile).

Transition metals are strong.

The transition metals are ***not very reactive***.
However, some do react slowly in air (oxygen) and water.
They form ***coloured compounds***. This has led to their use in patterns on pottery glazes.

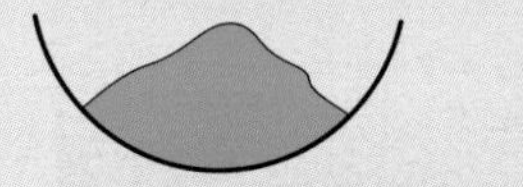

copper(II) chloride

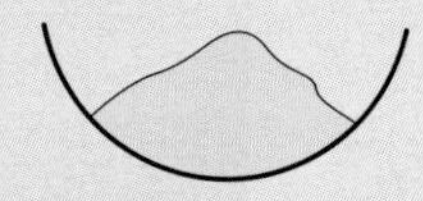

nickel chloride

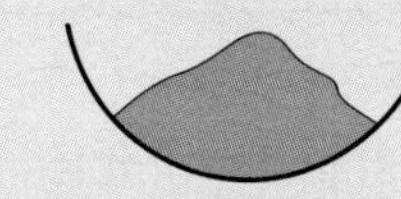
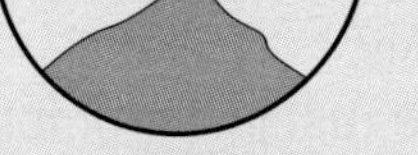

iron(III) chloride

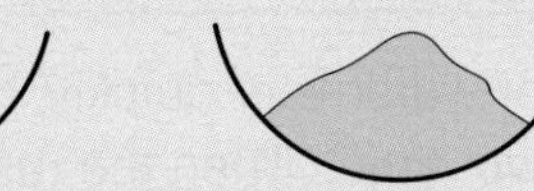

cobalt(II) chloride

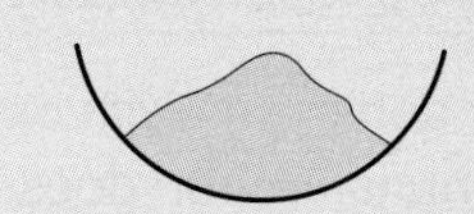

manganese(II) chloride

Transition metals are also important ***catalysts*** in industry.
(Catalysts make reactions go faster. They remain chemically unchanged themselves at the end of the reaction.) (See page 132.)
For example, iron is used in the making of ammonia. (See page 154.)
Platinum is one of the catalysts used to make nitric acid.
It is also used in catalytic converters in car exhausts. (See page 132.)
Catalytic converters help to reduce air pollution from cars

Questions

1 Copy and complete:

Over 75% of the chemical elements are ……
When the elements are sorted by atomic masses (with a few exceptions) we get the …… Table.

In Group 1 we find the …… metals.
In the central block we have the …… metals.

Comparing these metals, we see that the …… metals are very reactive. In water, they give off …… gas and form an …… solution of the metal ……. However, the metals from the central block, if they react at all, corrode …… in air and water.
The transition metals are useful in the chemical industry as …… to make reactions happen more ……

Most compounds of alkali metals are ……, whereas the compounds of transition metals are …… For example, sodium chloride is white and copper sulphate crystals are ……

The physical properties of these 2 sets of metals are also very …… For example, no solid blocks of a …… metal float in water, but the first 3 metals in Group …… do float – they have low ……
The transition metals are also ……, strong and tough.

2 Which of the transition metals is:

a) used in electrical wiring

b) the main metal found in steel

c) one of the catalysts used to make nitric acid

d) the silvery metal in a 50p coin

e) added to iron, along with nickel, to make stainless steel

f) given the symbol Ti.

3 Look at the densities of the alkali metals below:

alkali metal	Density (g/cm^3)
lithium	0.53
sodium	0.97
potassium	0.86
rubidium	1.53
caesium	1.88

a) Draw a bar chart to show the data in the table (use a computer if possible).

b) Which of the metals will float in water?

c) Why are the alkali metals stored in oil?

d) Describe what you would ***see*** when sodium is added to a bowl of water.

e) Write a word equation for the reaction of sodium with water.

f) When potassium reacts with chlorine it forms potassium chloride.
Write a word equation to show the reaction.

g) Predict the colour of potassium chloride.

h) Will potassium chloride be soluble in water?

i) Write a word equation for potassium reacting with oxygen.

j) The product formed in part i) is added to water. Do you think the solution formed will be acidic, neutral or alkaline?

4 Make a table to summarise the differences between the alkali metals and their compounds and the transition metals and their compounds.

5 The elements were first arranged in order of atomic mass in the 1800's.

a) What problem did they find when trying to place similar elements in the same columns? (See page 19.)

b) Name two elements that troubled them.

CHAPTER 3

3a Ores

We all take metals for granted, and many people have no idea where they come from.
Do you know where we get gold, iron or aluminium from?
In this chapter you will find out.

Most metals are found in nature as compounds with non-metals.
The metal and non-metal are chemically bonded together (often in oxides or sulphides).
If we find a lot of the metal compound in a rock, it is then worthwhile extracting the metal from it.
We call the rock a **metal ore**.

Rock salt is the raw material for making chlorine. We also extract sodium metal from this ore.

a) Rock salt is an ore containing sodium chloride.
Which metal do we extract from rock salt?

Look at the metal ores and the products we get from them below:

haematite

bauxite

iron (used to make steel)

steel cutlery

aluminium

aluminium cans

b) Which ore do we get iron from?

c) Which ore do we get aluminium from?

d) Find out the names of the main metal compounds in the iron ore and aluminium ore.

Metals that aren't reactive

Do you have a favourite metal? Lots of people like gold or silver jewellery.
Platinum, and sometimes copper, are also used for jewellery.

e) What are the most important properties for a jewellery metal?

We can find all of these metals in nature as the metals themselves. They exist as the elements – not bonded to non-metals (unlike most other metals). We say they are found **native**. This is because they are not very reactive. They have never reacted with the non-metals on Earth.

f) Which non-metal in the air do most other metals react with?

Look at the gold nugget opposite:

Gold is a rare metal. In the mines where it is found, you have to dig out tonnes of rock to get a tiny amount of gold.

g) Gold is mined in South Africa from veins of gold deep underground. Some people 'pan' for gold in rivers. How do you think that the gold gets into river beds?

h) Why do you think that gold is so expensive?

Copper and silver are more reactive than gold and platinum. We can still find them as native metals but they are also found as compounds. We then have to extract the metal from its compound. However, it is quite easy to get the metal back from the ore. For example, we find copper in an ore called chalcocite. We can ***extract the copper just by heating the ore***.

Panning for gold in a river.

Remind yourself!

1 Copy and complete:

Most metals occur in compounds, combined with ………… Some rocks contain enough of these compounds to ………… the metal from. These are called …………

2 Some metals are found native.

a) What does this mean?

b) What type of metals are found native?

c) Name 4 metals that can be found native.

d) Find out more about gold mining.

3b Reactivity Series

Are you wearing or carrying any metals that are not very reactive? Why is this important?
You know that other metals are very reactive.

We can put the metals in order of their reactivity.
It's like a league table of metals.
We call it the **Reactivity Series**.

a) Name two very reactive metals.
b) Name two metals that are not very reactive.

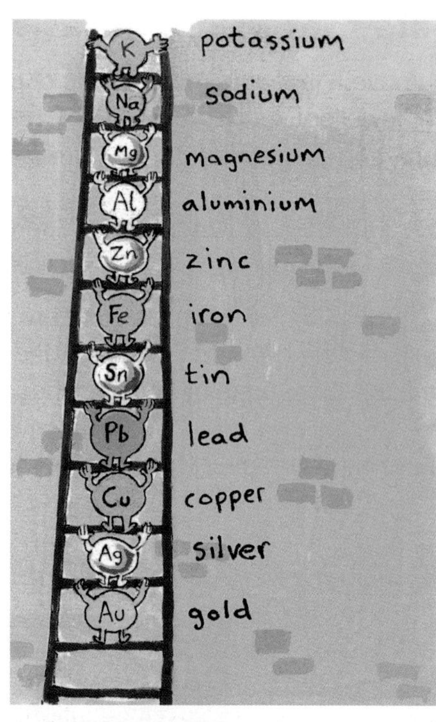

The Reactivity Series.

Look at the reactions of some metals in the table below:

<table>
<tr><th>Order of reactivity</th><th>Reaction when heated in air</th><th>Reaction with water</th><th>Reaction with dilute acid</th></tr>
<tr><td>potassium</td><td rowspan="8">burn brightly, forming oxide</td><td rowspan="4">fizz, giving off hydrogen; alkaline solutions (hydroxides) are formed</td><td rowspan="3">explode</td></tr>
<tr><td>sodium</td></tr>
<tr><td>lithium</td></tr>
<tr><td>calcium</td><td rowspan="5">fizz, giving off hydrogen</td></tr>
<tr><td>magnesium</td><td rowspan="4">react with steam, giving off hydrogen; the metal oxide is made</td></tr>
<tr><td>aluminium</td></tr>
<tr><td>zinc</td></tr>
<tr><td>iron</td></tr>
<tr><td>tin</td><td rowspan="3">oxide layer forms without burning</td><td rowspan="2">only a slight reaction with steam</td><td rowspan="2">react slowly with warm acid</td></tr>
<tr><td>lead</td></tr>
<tr><td>copper</td><td rowspan="4">no reaction, even with steam</td><td rowspan="4">no reaction</td></tr>
<tr><td>silver</td><td rowspan="3">no reaction</td></tr>
<tr><td>gold</td></tr>
<tr><td>platinum</td></tr>
</table>

c) Name a metal that reacts with steam, but not with cold water.
d) Name a metal that forms a layer of oxide when heated in air, but does not react with steam.
e) Why should we never add sodium to dilute acid?
f) Arrange the metals in the table into 3 league tables. Call them the Premiership, Division 1 and Division 2. Explain how you decided on each division.

Displacement reactions

As well as a league table, we can also put metals into competition with each other. (A bit like the F.A. Cup!) For example:

copper sulphate + zinc → zinc sulphate + copper

Zinc is added to copper sulphate.

This is called a **displacement reaction**.
The zinc is more reactive than the copper.
It displaces ('kicks out') the copper from the solution.
Copper is left as the metal and zinc goes into the solution.

Copper and zinc sulphate are formed.

> A more reactive metal **displaces** a less reactive metal from its compounds.

g) Write a word equation for magnesium reacting with copper sulphate solution. What would you see happen?

The reaction above happens in solution.
But you can also show displacement by heating metal and metal oxide powders together:

iron oxide + aluminium → aluminium oxide + iron

This reaction gets so hot that the iron formed melts.
This reaction is used to join tracks together on railways.
Look at the photo opposite:

Welding tracks using the 'thermit reaction'.

h) Explain how the workers in the photo are joining the rails together.

Aluminium and Iron both want the oxygen.

Aluminium wins the 'tug-of-war'.

Remind yourself!

1 Copy and complete:

The metals can be put in order of reactivity in the Reactivity ......
Highly reactive metals, such as ......, are put at the top and the least ...... metals at the ......

A metal higher up the list can ...... a ...... reactive metal from its compounds.

2 Which of these will react together?
If they do react, write a word equation.

a) copper oxide + zinc
b) magnesium oxide + lead
c) lead oxide + copper
d) silver oxide + copper
e) zinc + lead nitrate

3c Reducing metal oxides

Every time you switch on a light bulb, you can thank tungsten metal that it works!
We use tungsten to make the thin piece of wire inside a light bulb.
The metal itself does not occur as the element naturally.
We have to extract tungsten from its ore.

a) Do you think that tungsten melts easily?
How can you tell?

b) Why do we have to extract tungsten from its ore?

We use tungsten to make the filament in a light bulb.

We need to know a little more chemistry to explain how we get tungsten from its ore.
Tungsten is found in nature as tungsten oxide.
We have to remove the oxygen to be left with tungsten metal.

Removing oxygen from an oxide is called **reduction**.

You know about the Reactivity Series of metals.
But we can also include two useful non-metals in the Series.
Look at part of the Reactivity Series below:

magnesium
aluminium
CARBON
zinc
iron
tin
lead
HYDROGEN
copper
silver
gold

c) Which two metals in the list opposite are more reactive than carbon?

d) Can carbon 'take away' the oxygen from magnesium oxide?

e) Can carbon 'take away' the oxygen from zinc oxide?
Explain your answer.

Look at the reaction below:

Lead oxide + carbon → lead + carbon dioxide

The lead oxide has been **reduced** by the carbon (had its oxygen 'taken away').
The carbon is called a **reducing agent**.
Carbon is used to extract metals from ores.
It is cheap and there's lots of it available (we get it from coal).

Extracting tungsten metal

Tungsten fits in the Reactivity Series below both carbon and hydrogen.
So we could use carbon to reduce tungsten oxide.
Unfortunately the carbon reacts with the tungsten formed, making it brittle. But we can use hydrogen:

tungsten oxide + hydrogen → tungsten + hydrogen oxide (steam)

f) What has been reduced in the equation above?
g) What is the reducing agent in the equation above?

Look at some of the uses of tungsten metal below:

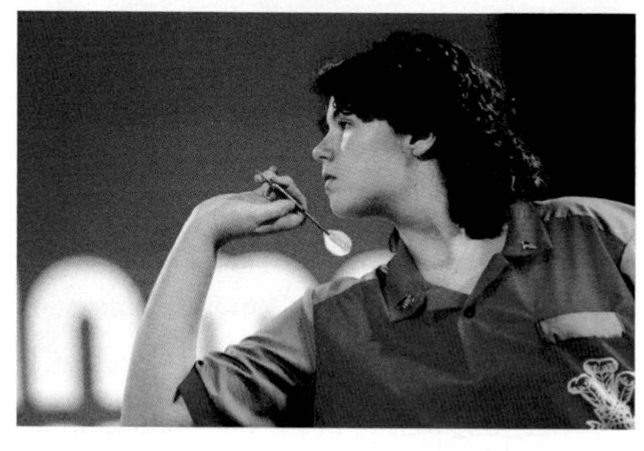

The shafts on these darts can be made very thin using tungsten.

Tungsten steel is used as the cutting edge on a lathe.

Tungsten alloys are used to make some golf clubs.

h) Which properties of tungsten metal make it good for the uses shown above?

Remind yourself!

1 Copy and complete:

Carbon can be used as a ......
agent to extract metals from oxides.
We say that the metal oxide has been
...... in the reaction.
...... gas is used to extract tungsten
metal from its ore (tungsten ......).

2 a) Finish the equation below:

carbon + zinc oxide → ...... + carbon dioxide

b) The formula of zinc oxide is ZnO and we can show carbon as C.
Write a balanced symbol equation for the reaction in part a).

3d The Blast Furnace

The most important metal in modern life is steel.
Steel is usually over 97% iron.
So before we can make steel, we need to extract iron.
We extract the iron from its ore called **haematite**.
Haematite is **iron(III) oxide** (Fe_2O_3).

A Blast Furnace is over 50 metres high.

a) Look back to the list on page 30:
Can we use carbon to extract iron from its ore?
Why?

The haematite ore is imported in ships.
From the port, it is taken to a **Blast Furnace**.

The iron(III) oxide is mixed with **coke**, a cheap form of carbon.
Limestone is added and it's all fed into the top of the furnace.
Hot air is blasted into the bottom of the furnace.
Look at the diagram below:

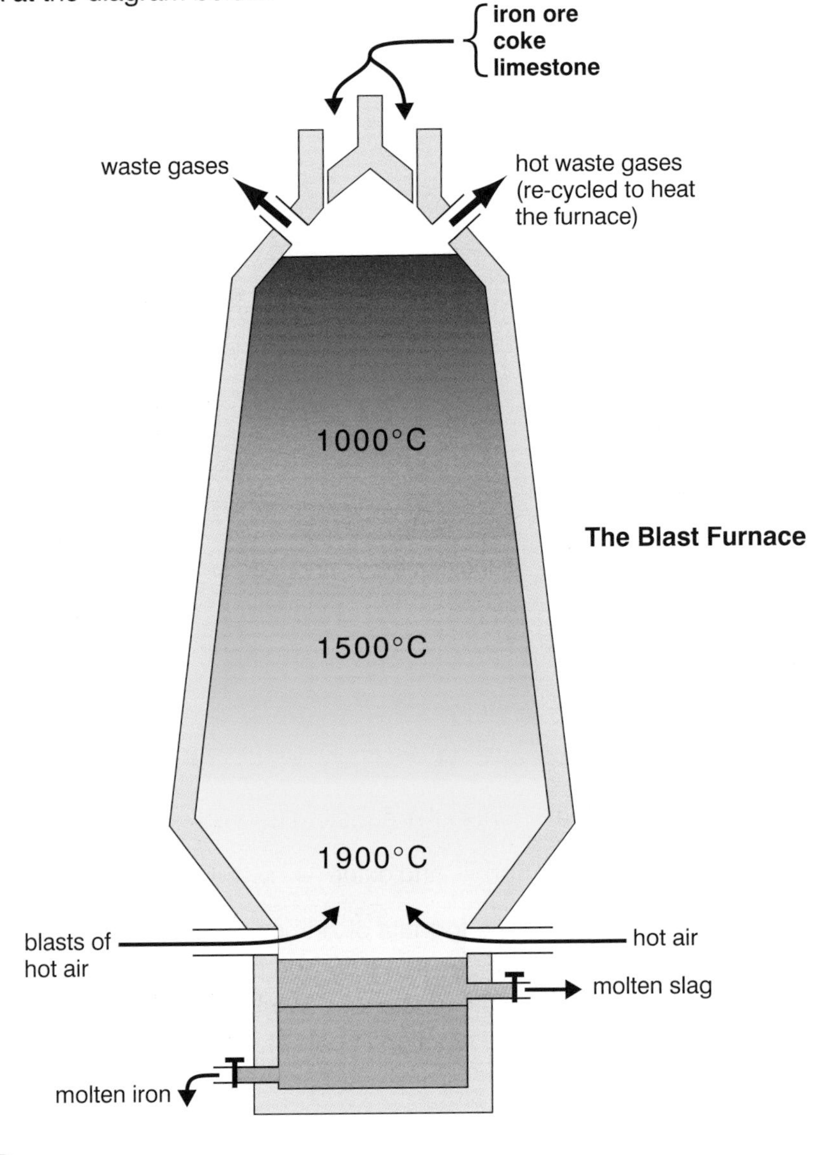

The Blast Furnace

Some of the iron(III) oxide is ***reduced*** by carbon:

Iron(III) oxide + carbon → iron + carbon dioxide

$2\ Fe_2O_3(s) + 3\ C(s) \rightarrow 4\ Fe(l) + 3\ CO_2(g)$

b) What is the temperature in the middle of the Blast Furnace?

c) Is the iron formed as a solid, a liquid or a gas? Why?

d) Where does the iron collect in the furnace ?

But **carbon monoxide gas *is the main reducing agent*** in the Blast Furnace. Let's see how it forms:
The coke starts to burn, like the charcoal in a barbecue:

carbon + oxygen → carbon dioxide

$C(s) + O_2(g) \rightarrow CO_2(g)$

e) Where does the oxygen in the furnace come from?

Then the carbon dioxide in the furnace reacts with more hot coke (carbon):

carbon dioxide + carbon → carbon monoxide

$CO_2(g) + C(g) \rightarrow 2\ CO(g)$

f) What gets reduced (has oxygen taken away) in the equation above?

The carbon monoxide can mix well with the iron(III) oxide.
It is a gas, so it can get into any small gaps.
It reduces the iron(III) oxide as shown below:

iron(III) oxide + carbon monoxide → iron + carbon dioxide

$Fe_2O_3(s) + 3\ CO(g) \rightarrow 2\ Fe(l) + 3\ CO_2(g)$

The ***limestone*** in the furnace ***gets rid of acidic impurities*** (rocky bits).
It breaks down then reacts with them, forming a **slag**. This floats on top of the molten iron at the bottom of the furnace.
The slag is tapped off and cooled. It is then used in making buildings and roads.

State symbols

In balanced chemical equations you will sometimes see: (s), (l), (g) or (aq) after a formula.
These are state symbols, and stand for:

(s) = solid
(l) = liquid
(g) = gas
(aq) = dissolved in water

Remind yourself!

1 Copy and complete:

Iron is extracted from its ore, called ......, in a ...... Furnace. The ore is mixed with ...... (a form of carbon) and ......
Hot ...... is blasted into the furnace and molten ...... is tapped from the bottom.

2 a) What reduces most of the iron(III) oxide in the Blast Furnace.

b) How is this reducing agent made in the furnace?

c) How are the acidic impurities (rocky and sandy bits) removed from the furnace?

d) What is slag used for?

3e Turning iron into steel

Imagine if you had a slight bump in your car and it cracked in half! That would surprise you, but it could happen if your car was made from iron straight from the Blast Furnace.

Iron from a Blast Furnace has carbon, and other impurities, mixed in with it.
This makes it hard, but brittle. At this stage, it is called pig iron. (A pig is the name given to the moulds the molten iron runs into from the bottom of the furnace.)

We turn this form of iron into the more useful metal, **steel**.

a) Why do you think that steel is a more useful metal than pig iron?

Turning iron into steel involves two steps:
- removing most of the impurities
- adding small amounts of other metals (**alloying**).

b) Name one impurity that we remove from pig iron.

Step 1

Most of our steel is made using the **Basic Oxygen Process**. This gets rid of carbon from the iron by reacting it with oxygen. The carbon dioxide formed floats off as a gas. Look at the diagram below:

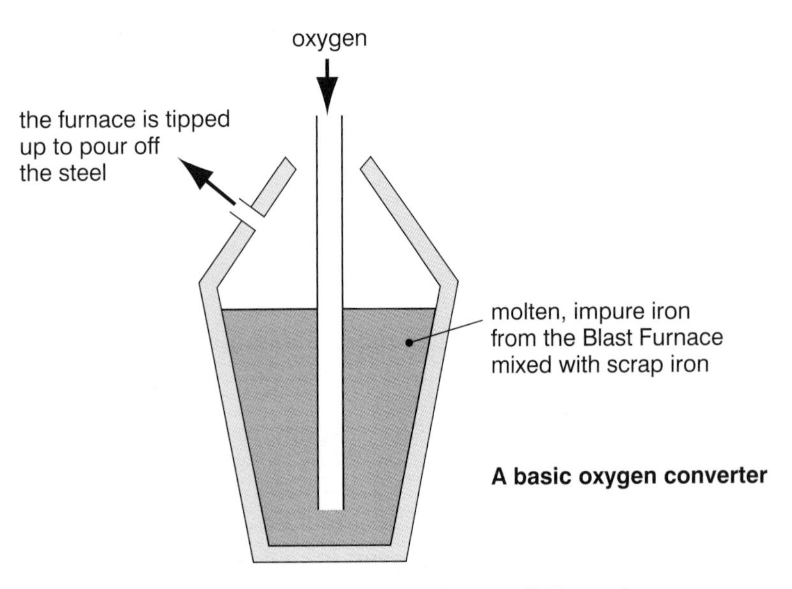

A basic oxygen converter

Iron is turned into steel. The oxygen burns off the carbon and other non-metal impurities.

Making steel
In order to make steel less brittle,
Its carbon we must try to whittle.
Blow oxygen through
To remove CO_2
Then add metals – but only a little!

c) In the Basic Oxygen Process, we change pig iron and which other type of iron into steel?
d) Write a word and symbol equation for the reaction that removes carbon in the Process.
e) Why it is important to ***recycle*** iron and steel.

Car bodies are made from mild steel.

The steel made in the Process still has some carbon left in it. The actual amount changes the hardness of the steel made. For example, mild steel, used to make car bodies, contains 0.2% carbon.

f) Pig iron contains 3.5% carbon.
What percentage of carbon is removed when we make mild steel?
g) How do the properties of mild steel differ from pig iron?

Step 2

Special steels can be made by adding small amounts of other metals. These are called **alloy steels**.

An **alloy** is a mixture of metals.

For example, a very hard and tough steel is made using a little ***tungsten***.
Stainless steel is made by adding a little ***nickel and chromium***.
It doesn't rust.

h) Suggest a use for:
i) tungsten steel
ii) stainless steel.

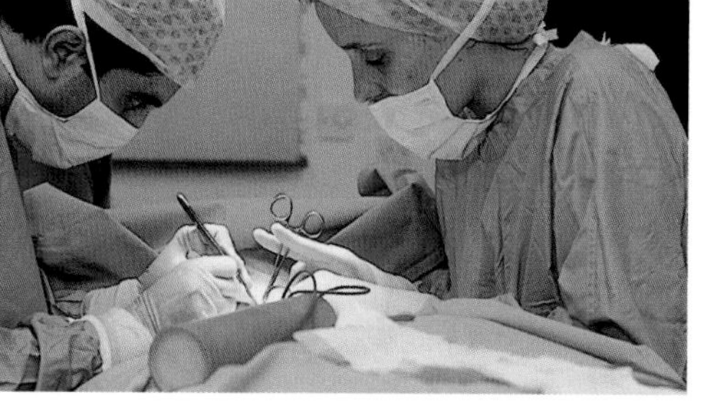
Surgeons use stainless steel instruments.

Remind yourself!

1 Copy and complete:

...... from the Blast Furnace is not pure. Its main impurity is ......

To make steel we have to ...... the impurities. ...... gas is piped in to the molten mixture and much of the carbon forms .... gas.

We can add small amounts of other metals to form ...... steels with improved properties.

2 a) What is an alloy?
b) How do you think that the mixture is formed in an alloy?
c) Which metal do we add to steel to make a very hard, tough alloy?
d) Which metals can we add to steel to make stainless steel?
e) Design a poster persuading people to recycle iron and steel.

3f Rusting

The problem with iron (and most steel) is rusting.
Rust costs us millions of pounds every year.
Have you ever had anything spoilt because it went rusty?
If you haven't used your bike for a while, sometimes the chain rusts up.

The iron in mild steel corrodes to form rust.

a) How do we protect a bicycle chain from rusting?
b) How does this stop the iron from rusting?

Look at the results of an experiment below:

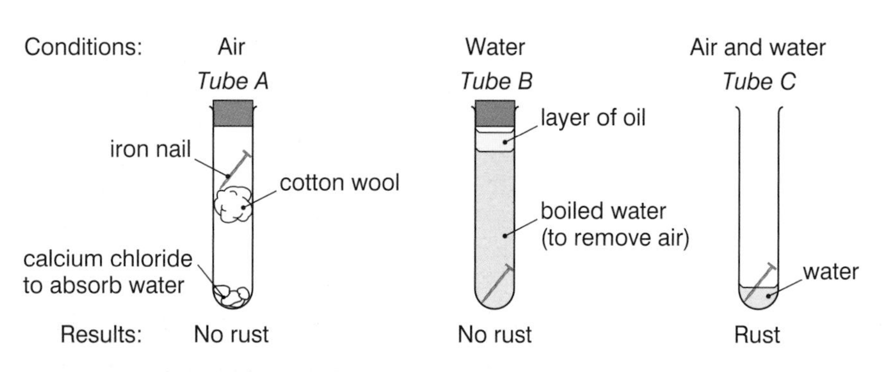

In the experiment above:

c) How do we remove dissolved air from the water?
d) How do we make sure no new air gets back into the water?
e) How do we get rid of the water vapour from the air?
f) Which tube contains i) only water, ii) only air, iii) both air and water?

This experiment shows that:

Both air (oxygen) and water are needed for iron to rust.

g) Which gas in the air reacts with the iron in rusting?
h) Sort out these letters to find another word for rusting:

r o n o r
i C s o

We can think of rust as a type of iron oxide with water trapped in its structure. Some metals corrode but are protected by the layer of oxide on its surface. Unfortunately, rust crumbles away and exposes fresh iron to air and water. So rust weakens iron and steel.

Rusting
In order for iron to rust,
Both air and water's a must.
Air alone won't do
Without water there too,
So protect it or get a brown crust!

Protection against rust

The obvious way to protect iron and steel is to keep the air and water away from it.
You ***form a barrier*** on the surface of the iron with:

- oil or grease
- paint
- plastic, or
- another metal.

i) Name one object protected by each method above.

Tin cans are actually steel cans coated by a very thin layer of tin. The tin keeps air and water from the iron.
But what happens when the tin is scratched?
Does the can rust?
People can get food poisoning if a can rusts and lets bacteria in.

j) Look back to the Reactivity Series on page 28.
Is tin more reactive or less reactive than iron?

A better way to protect iron or steel is to
coat it with a more reactive metal.
Have you noticed the shiny metal surfaces on some wheelie-bins?
We use zinc to coat the steel because a bin is likely to get knocked about. But even when scratched, the zinc still protects the iron. Remember that zinc is more reactive than iron. So the air and water will attack the zinc rather than the iron.
We call this **sacrificial protection**.

*This bin has been **galvanised** (coated with a thin layer of zinc).*

k) Do you think that magnesium would be a good metal to protect iron? Explain why.

l) Why do you think this method is called ***sacrificial*** protection?

Preventing rust
Whenever you come across rust,
Your iron will turn into dust.
Air and water will spoil it
If you don't try to foil it;
Sacrificial protection's a must!

Another good way to stop iron corroding is to make **stainless steel**.
We looked at steel alloys on page 35.
Adding small amounts of ***nickel*** and ***chromium*** to molten steel form this rust-proof alloy. However, it is expensive.
That's why we still use cheaper, but less effective, methods like painting.

Remind yourself!

1 Copy and complete:

For iron or steel to rust, it must have both ...... and ...... to react with.
Some methods of protecting it, such as ...... cans, form a barrier around the iron.
However, coating it with a ...... reactive metal, such as ......, protects the iron even when scratched.

2 a) Why aren't ships built from stainless steel?

b) Why is oil or grease useful for protecting some things?

c) Underground iron pipes and the hulls of ships can be protected against rust by attaching blocks of ***magnesium***. What do we call this type of protection? How does it work?

3g Extracting highly reactive metals

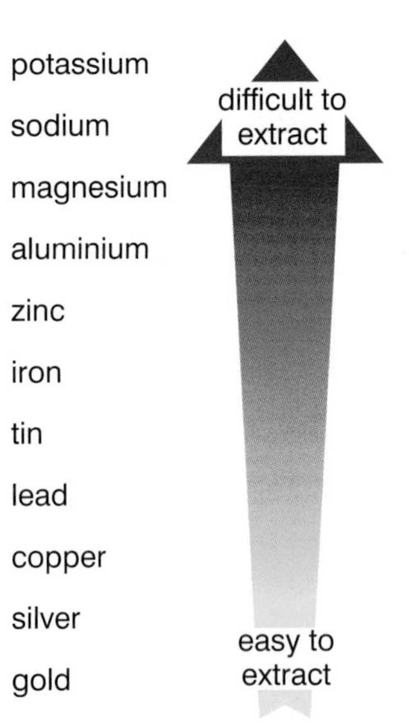

In this chapter we have seen how the metals at the bottom of the Reactivity Series are easy to extract.
Some are found as the metal itself in nature. If they are found in a compound, you can often just heat the ore to get the metal.

Metals of medium reactivity are a little harder to extract.
These are often found as oxides in their ores.
Some occur as sulphides (for example, lead sulphide, PbS).
These sulphides can be changed to oxides by heating.

Then we have to do some neat chemistry to remove the oxygen from the oxides. This often involves heating with carbon.

The metals at the top of the Reactivity Series are even more difficult to 'win' from their ores.

To extract reactive metals we have to:

1. separate the metal compound from the rest of the ore,
2. melt the metal compound, then
3. pass electricity through the molten compound.

You can imagine that this is an expensive process.
It takes a lot of energy to melt the metal compound.
Then there is all the electricity needed to finally get the metal.

You can read more about this (called **electrolysis**) in the next chapter.

Reactive metals are more difficult to extract than unreactive metals.

Reactive metals form stable compounds. This makes them difficult to extract!

It takes a lot of energy to split up compounds of reactive metals, like sodium.

Summary

Metals are found in the Earth's crust as pure metals (for example, gold) or as metal compounds.
Ores are rocks that contain enough of a metal or its compound to make it worthwhile extracting.

We can predict how to **extract** a metal from its position in the Reactivity Series.
The highly reactive metals are difficult to extract.
We use **electrolysis** to extract these metals, such as sodium or aluminium.
The metals of medium reactivity can be extracted by **reducing** their oxides using carbon.
The main example of this is iron, extracted in a Blast Furnace.

The more reactive a metal, the more difficult it is to extract.

Questions

1 Copy and complete:

An ...... is rock that contains a metal, such as ......, or a metal compound, for example, iron(III) ...... Metals that are ...... are easier to extract than reactive metals. The reactive metals are extracted using ......
whereas iron is extracted by ...... its oxide.

2 Copy and complete:

We extract ...... in a Blast Furnace. Haematite (iron ......), coke and ...... are fed into the top of the furnace continuously.
The main ...... agent in the Blast Furnace is carbon ...... gas. The metal is tapped off as a liquid from the ...... of the furnace. Molten ...... floats on top of the ...... and is tapped off separately.

3 a) What is the Reactivity Series?

b) Which metals in the Reactivity Series are most difficult to extract?

c) Use the Reactivity Series to predict which of these will react together:
i) zinc oxide and iron
ii) lead oxide and iron
iii) copper sulphate and zinc
iv) copper sulphate and silver.

d) Write word equations for the reactions you predicted would happen in part c).

e) What do we call the type of reaction in part d)?

4 a) Complete this word equation:

iron oxide + aluminium → ...... + ...

b) Explain the reaction in part a).

c) How is this reaction used on the railways?

d) Choose another metal that would react even more violently with iron oxide than aluminium.

5 Look at the table below:

Metal	Known since
potassium	1807
zinc	before 1500 in India and China
gold	ancient civilisations

a) Which metal in the table was discovered most recently?

b) Explain why each metal was discovered at the time shown in the table.

6 We extract titanium metal using sodium or magnesium metal. The last step in the process is:

titanium chloride + sodium → titanium + sodium chloride

a) Explain what happens in this reaction.

b) Write a word equation for the same reaction but using magnesium instead of sodium.

c) Write a word equation to show how tungsten metal is extracted from its oxide by hydrogen.

7 Look at the photo below:

a) The pier has magnesium bolted to its iron legs. Explain how this stops the pier from rusting.

b) What do we call this method of stopping iron rusting?

c) Which metal is used to coat a can of baked beans? Why isn't this as effective at stopping rust as the method in part a)?

So why do we use a less reactive metal in food cans?

CHAPTER 4 Metals and Electrolysis

4a Electrolysis

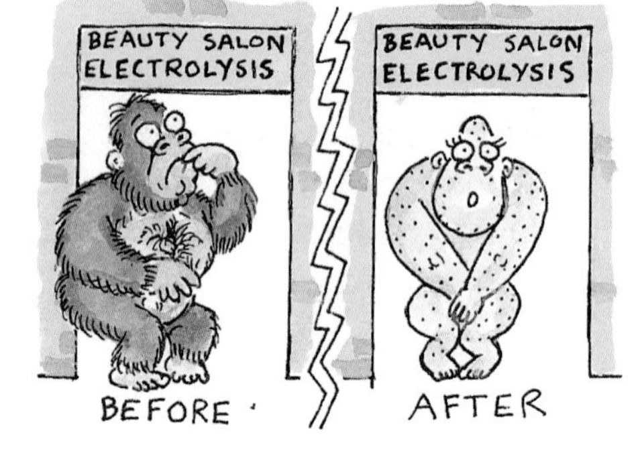

You might have come across the word 'electrolysis' in adverts for beauty treatments. It is used to remove unwanted hair. But in this chapter, we will see how electrolysis is important in giving us aluminium and copper.

a) Give one use of aluminium and one use of copper.

We find that some substances are broken down when electricity is passed through them. These substances have to be melted or dissolved in water.

Electrolysis is the breakdown of a substance by electricity.

b) Sort out these letters to find another word for 'breakdown'.

c D e n p m o
o s t i o i

Look at the experiment opposite:
Lead bromide is being ***electrolysed*** (broken down by electricity).

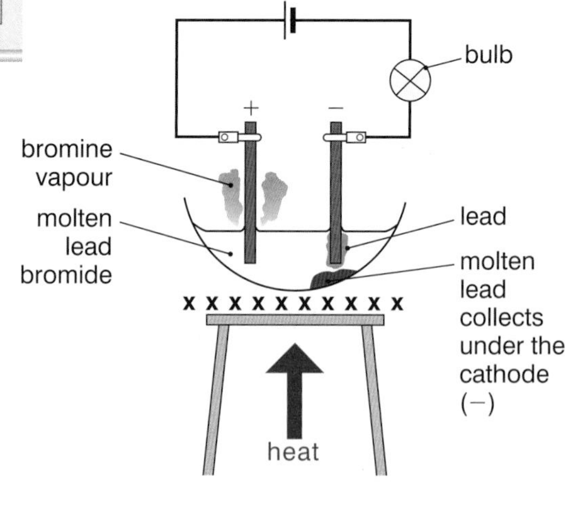

The electrolysis only starts when the lead bromide melts. No electricity flows when it is solid.

c) At which electrode (+ or –) does the lead form?
d) What forms at the other electrode?
e) Why would this experiment have to be done in a fume-cupboard?

Explaining electrolysis

Substances that can be electrolysed are usually compounds of metals and non-metals.

f) In lead bromide, name the metal and non-metal elements.

g) Write a word equation to show what happens when lead bromide is broken down.

Ions carry their charge to the electrode.

You've probably heard the saying 'opposites attract'.
That's not only true in romance!
In electrolysis opposite charges attract.
The lead bromide is made up of ***charged particles*** called **ions**.
It contains lead ions and bromide ions.

h) Look back at the experiment on the last page:
Try to work out the charge (positive or negative) on a lead ion.
Explain your answer.

Metal ions are always positively charged.

So metal ions always go to the negative electrode during electrolysis.
We call the ***negative electrode*** the **cathode**.
The ***positive electrode*** is called the **anode**.
During electrolysis:

Metals are always formed at the negative electrode (the cathode).

i) Why is bromine formed at the positive electrode (the anode)?

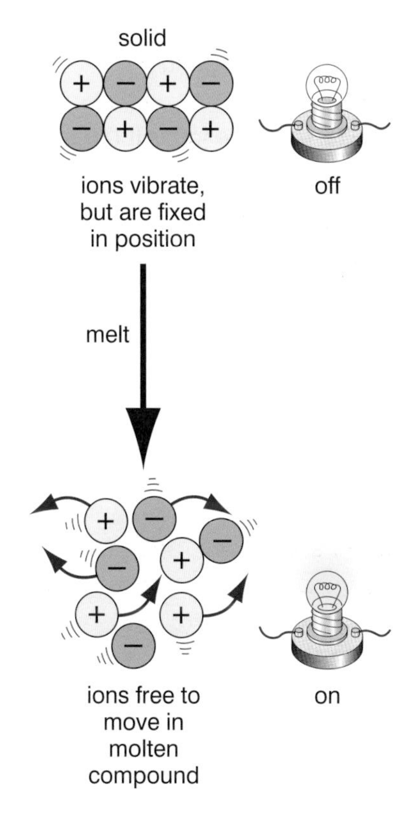

Notice that the ***electrolysis only happens once the lead bromide melts***.
This is because the ions have to be able to move to the electrodes.
Remember that in a solid, particles can't move about.
They vibrate but are stuck in position.
Once we heat the solid until it melts, the particles can slip and slide over each other.
So once lead bromide melts, the ***ions become free to move*** to the electrodes. Then the electrolysis happens.

Remind yourself!

1 Copy and complete:

When we break down a substance using electricity, it is called ……
Metal ions are …… charged and are attracted to the …… electrode (the cathode).

2 a) You electrolyse molten magnesium chloride. What would be formed:

i) at the cathode ii) at the anode?

b) Explain why we have to melt the magnesium chloride before it can be electrolysed.

4b Extracting aluminium

Aluminium is the most important of the reactive metals that we extract by electrolysis. You probably drank your last fizzy drink from an aluminium can. Did you recycle your can?

Aluminium cans.

Look back to page 30 to see where aluminium and carbon are placed in the Reactivity Series:

a) Which is higher in the Series, aluminium or carbon?
b) Explain why we can't use carbon to reduce aluminium oxide.

We find aluminium in an ore called **bauxite**.
You have seen a photo of the ore on page 26.
The ore contains aluminium oxide.
Our job is to get rid of the oxygen and leave aluminium metal.
In other words, we have to reduce the aluminium oxide.
But because the aluminium is reactive this isn't easy.
Remember that we have to melt the metal compound, then pass electricity through it.

Opencast mining of bauxite.

c) Why do we have to melt the metal compound before electrolysis?

In this case, the metal compound is aluminium oxide.
It is a white compound. So you can see that it is clearly mixed with other coloured things in its ore, bauxite.
The first step is to separate the aluminium oxide from the mixture.
The brown waste material is stored in lagoons.
Look at the photo below:

d) How does the mining of bauxite affect our environment?
e) How can we help to ease the problems?

Once you have the aluminium oxide you can electrolyse it.
This is done in a factory using lots of the cells shown below:

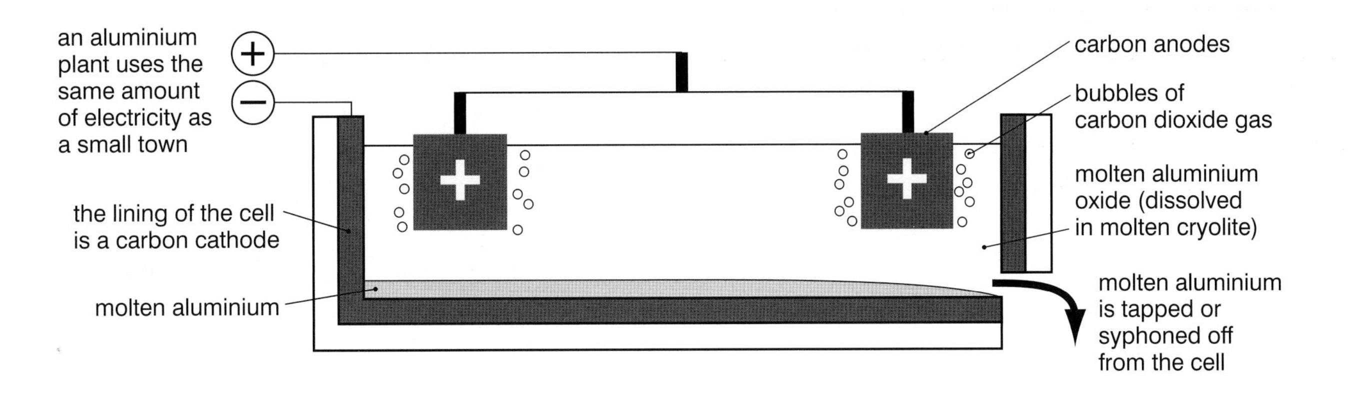

Look at the diagram above:

f) What are the electrodes made from?

g) Which electrode (+ or –) is aluminium formed at?

h) What forms at the other electrode?

The aluminium oxide is mixed with, and dissolves in, **molten cryolite** in the cell. This lowers its very high melting point.
You can melt the aluminium oxide at a lower temperature and save money on heating!

At the ***negative*** electrode (cathode) we get ***aluminium***.
At the positive electrode (anode) we get oxygen gas.

The oxygen gas formed at the positive electrode reacts with the hot carbon anodes. They burn away, forming ***carbon dioxide*** gas.
So we have to replace the anodes regularly.

i) Write a word equation for the reaction of oxygen with carbon.

j) Why do we use carbon electrodes instead of using an electrode that wouldn't react with the oxygen, such as platinum?

Remind yourself!

1 Copy and complete:

Aluminium is mined from an ore called ……
The aluminium …… is separated from the rest of the ore.
It is then electrolysed in a cell with electrodes made from …… The aluminium …… has its melting point lowered by mixing it with ……

2 Explain these things about the extraction of aluminium:

a) We don't use a metal, such as sodium, to displace aluminium in the process.

b) Aluminium plants are often found near hydro-electric power stations.

c) We replace the anodes frequently.

4c Uses of aluminium

Have you wondered why aluminium is such a useful metal when it seems so reactive?
Look at its position in the Reactivity Series below:

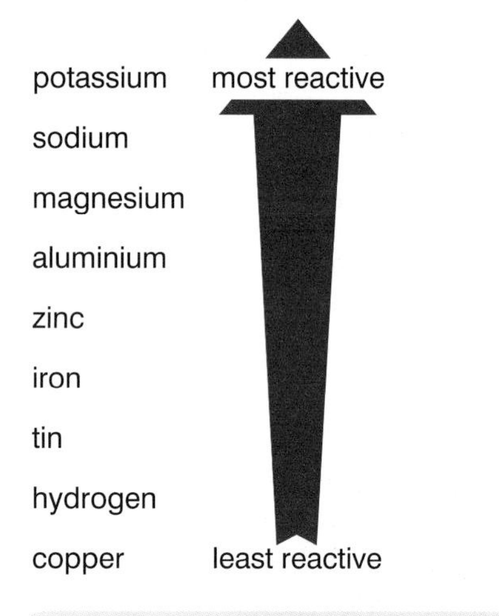

a) What would happen to the acid in a fizzy drink if the can was made from zinc or magnesium?

But aluminium cans don't corrode. They don't react with air or water (or even acid). So why is that?
It doesn't seem to make sense from the Reactivity Series.
Aluminium can even be used to make window frames.
These last for many years without needing to be replaced.

b) Why are aluminium windows sometimes replaced even though they haven't corroded?

The answer to aluminium's puzzling lack of reactivity is its coating.
Not a coating we put on the metal to protect it, as we do with iron or steel. There is no need for that!
Aluminium reacts in air to form a ***coating of aluminium oxide***.

This oxide layer is tough. It acts as a barrier on the surface of the aluminium. Air or water just can't get at the aluminium underneath. So it doesn't react (corrode) any further.

The frame of this greenhouse is made of aluminium.

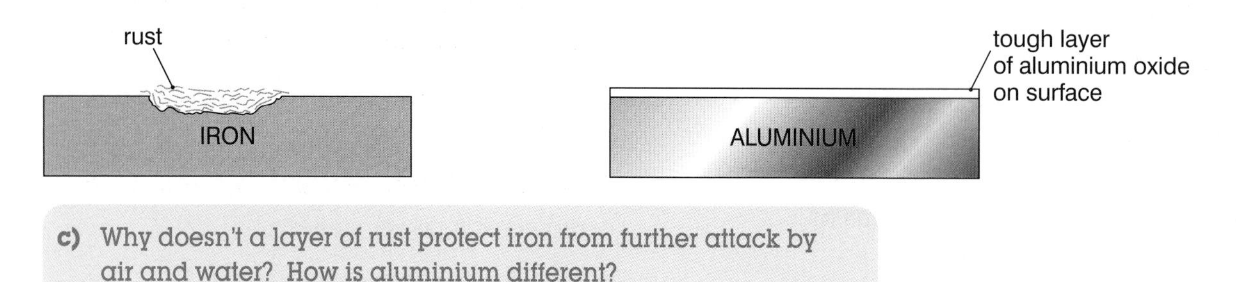

c) Why doesn't a layer of rust protect iron from further attack by air and water? How is aluminium different?

This means we can use aluminium for all sorts of structures that need to be weather-proof. It has some excellent properties, not least its ***low density***. Would you rather cycle a steel framed bicycle up a steep hill or an aluminium one? Why? However, you might change your mind if you crashed your bike. Which do you think is the stronger of the two metals?

We can ***strengthen aluminium by alloying*** it with other metals. For example, an alloy of aluminium mixed with a little magnesium and copper is made. It is used to make aeroplanes.

d) Why is aluminium an excellent metal for making passenger jets?

So how does alloying make a metal stronger?
Look at the diagram below:

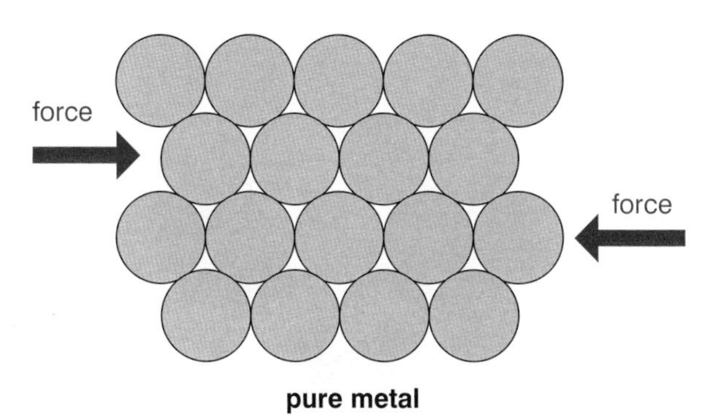

pure metal

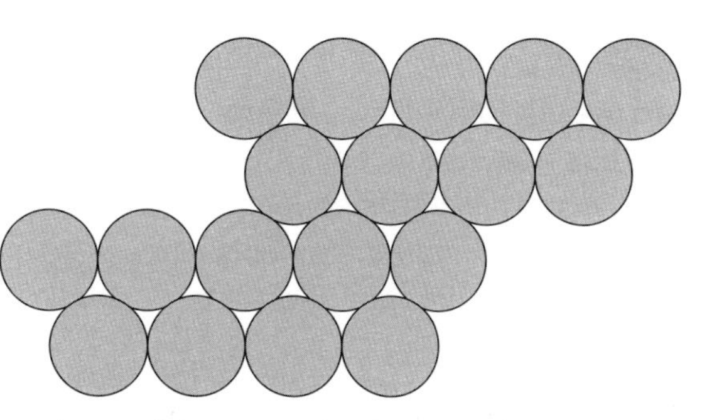

layers slide over each other easily in a pure metal

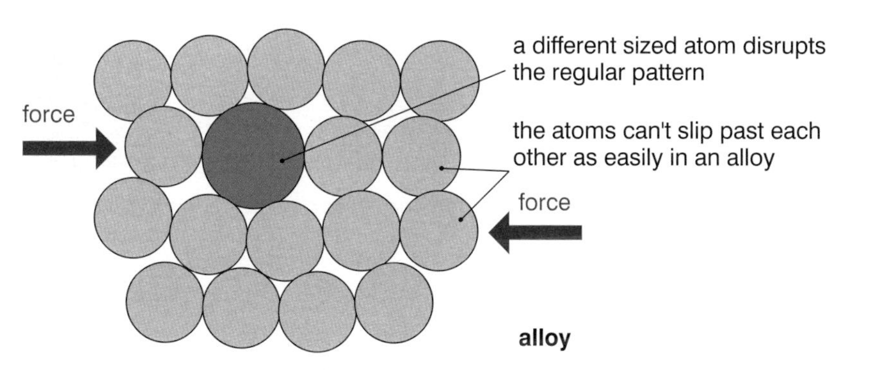

alloy

With different sized atoms mixed in, the layers can't slip past each other so smoothly. It's a bit like little stones jamming a door!

Remind yourself!

1 Copy and complete:

Aluminium is not as ...... as we would expect from its place in the ...... Series.
It is protected by a tough ...... of aluminium ...... on its surface.
We can make aluminium harder, ... or stiffer by adding other metals to make ......s.

2 a) Which metals do we add to aluminium to make an alloy used in aeroplanes?

b) Why do we need to use aluminium alloys for this job?

c) Explain how alloying makes a metal stronger.

d) List some other alloys.
Which metals are used to make them?

4d Purifying copper

When was the last time you used copper metal?
You might be thinking of the copper alloys used in coins.
But did you think of the last time you turned on a tap?
The water most probably arrived at the tap through copper pipes.
Or when did you last switch on the lights?
The electricity will have flowed through copper wires.

We use copper for water pipes.

a) What properties of copper make it good for the uses above?

But why is copper in this chapter on 'Metals and electrolysis'?
Surely it's the reactive metals that are extracted by electrolysis.
And we know how ***unreactive*** copper is!
In fact, electrolysis is used to get ***pure copper***.

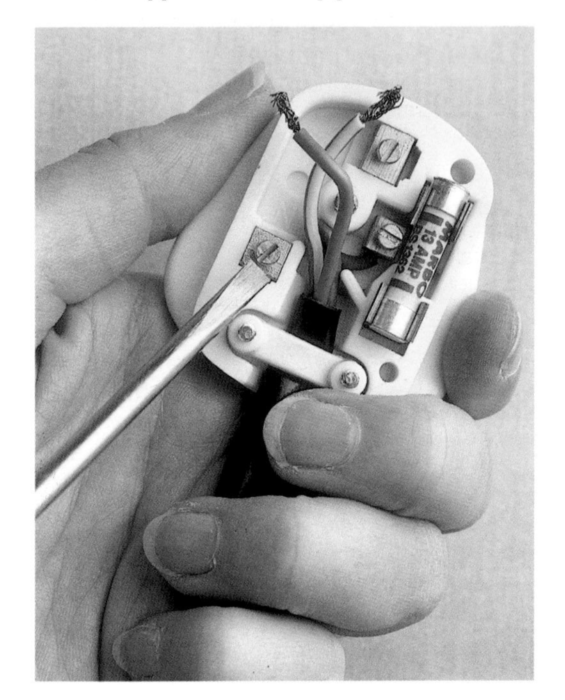

We use copper for electrical wiring.

We need very pure copper for electrical wiring.
But the copper that has been extracted from its ore has too many impurities. This affects copper's ability to conduct electricity. So the impure copper is placed in cells like the one shown on the next page.

b) Why do we need pure copper for electrical wiring?

c) Look back to page 41.
Do you think the charge on a copper ion is positive or negative?

Look at the electrolysis experiment below:

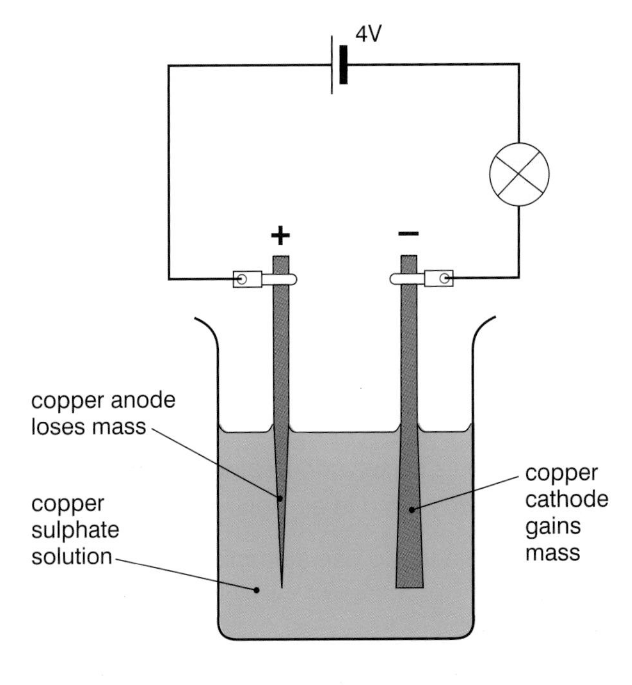

We find that the:

loss in mass at the anode (+) = gain in mass at the cathode (–)

d) What happens to the copper anode (+)?

e) What happens to the copper cathode (–)?

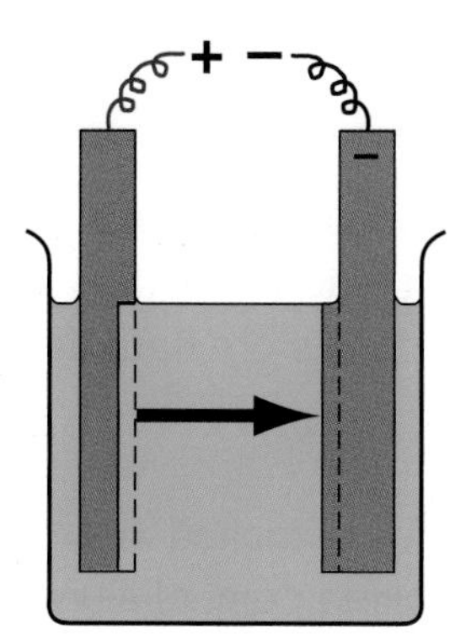

The anode's loss is the cathode's gain.

In industry we use cells like the one below:

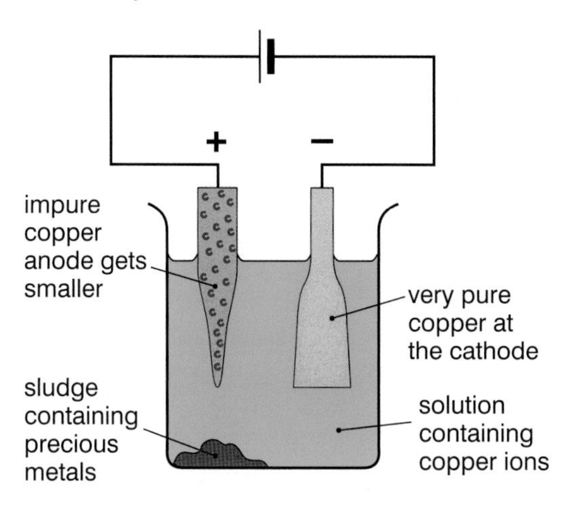

Many of these cells operate at the same time in industry.
The cathodes are removed after about two weeks.

The ***anode (+) is made of the impure copper*** from the ore.
The ***cathode (–) is a piece of pure copper***.
The solution they dip into contains copper ions. These are positively charged.

f) What happens to the impurities?

g) Do you think the impurities are just thrown away? Why?

The cathodes are replaced about every two weeks.

h) Why do we remove the cathodes regularly?

Remind yourself!

1 Copy and complete:

Copper is easy to extract from its ...... but we need very ...... copper for use in electrical ...... The impure copper is made the ...... in a cell, with ...... copper forming the ...... The solution contains copper ...... The ...... grows bigger and is replaced about every ...... weeks.

2 We can use electrolysis to plate metals with a thin layer of a different metal. This is called electroplating.

a) You want to plate a steel fork with nickel. What would you use as the cathode?

b) Which metal would you use as the anode?

c) Draw a diagram of your circuit.

Summary

Electrolysis is the breakdown of a substance by electricity.
It is used to extract reactive metals, such as **aluminium**.
Aluminium is found in its ore **bauxite**, which contains aluminium oxide.

The aluminium oxide is melted in a mixture with molten cryolite.
This lowers the melting point of aluminium oxide.
The electrodes are made of carbon.
Aluminium forms at the cathode (–).
Carbon dioxide is given off from the anode (+).
(The oxygen from the aluminium oxide reacts with the carbon anode to make carbon dioxide.)
So the anode gets burned away and has to be replaced often.

The aluminium made is a very useful metal.
It does not corrode because it is covered in a tough layer of aluminium oxide.

We can make aluminium stronger by forming **alloys** by mixing in small amounts of other metals, such as magnesium.
These alloys can be made much harder and stronger than pure aluminium.

Copper is ***purified*** by electrolysis.
The anode is the impure copper. The cathode is pure copper.
They dip into a solution containing copper ions which are positively charged.
The cathode gets larger as pure copper builds up on it.
Cathodes are taken away and replaced every couple of weeks.

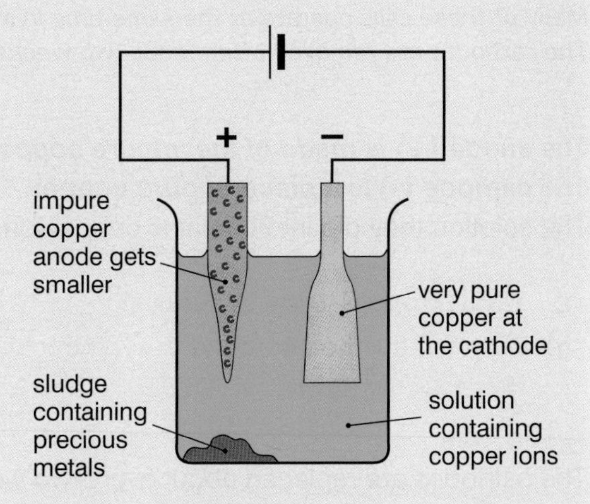

Questions

1 Copy and complete:

We use ...... to extract highly reactive metals. For example, ... has to be extracted from its ore called bauxite. Aluminium ...... is separated from the ore, then melted in a mixture with ...... (to ...... its melting point). In the cell, the electrodes are made from ...... Molten ...... forms at the cathode (the ...... electrode). Oxygen produced at the anode reacts with the carbon electrode and forms carbon ...... gas. This means that the anodes have to be ...... frequently.

Aluminium metal is ...... against corrosion by a layer of aluminium ...... The metal is made stronger, harder and stiffer by making ......

Copper is ...... using electrolysis. The impure copper is made the ...... electrode and a pure ...... cathode is used in a solution of copper ......

2 Look at the diagram below:

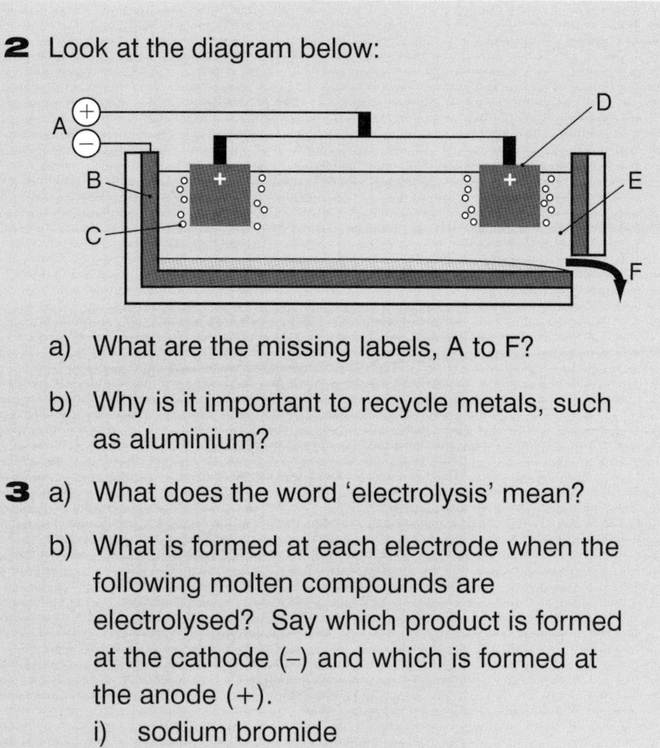

a) What are the missing labels, A to F?

b) Why is it important to recycle metals, such as aluminium?

3 a) What does the word 'electrolysis' mean?

b) What is formed at each electrode when the following molten compounds are electrolysed? Say which product is formed at the cathode (–) and which is formed at the anode (+).
   i) sodium bromide
   ii) potassium oxide
   iii) magnesium chloride
   iv) lithium iodide

c) Why do we have to melt the compounds in part b) before we can electrolyse them?

4 An alloy of aluminium is called duralumin. It is made by adding a little copper and magnesium to aluminium.
Duralumin is used to make aeroplanes.

a) Why are the other metals added to the aluminium?

b) Draw a diagram of the atoms in a pure metal and in an alloy. Use it to show how alloying a pure metal changes its properties.

5 a) Draw a diagram to show how we get pure copper from impure copper in industry.

b) In an experiment in the lab, a student electrolysed copper sulphate solution. She used copper electrodes and weighed them before and after the experiment.

The positive electrode lost 2.0 g.
What happened to the negative electrode?

c) If we left the current on, why would the electrolysis stop eventually?

6 Sodium is a reactive metal extracted by electrolysis in a Down's Cell:

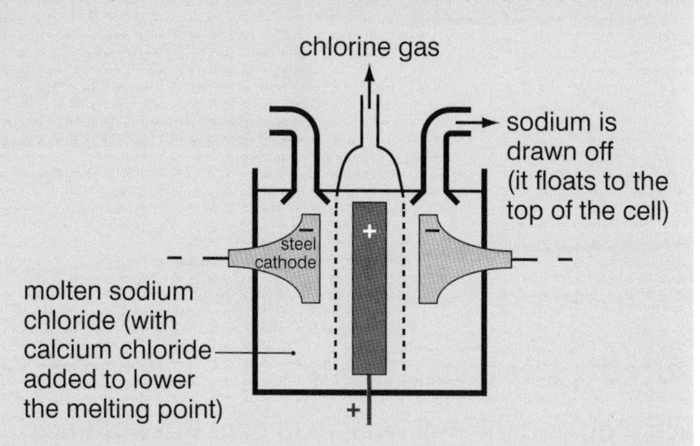

The Down's Cell for extracting sodium.

a) Which compound do we extract sodium from?

b) Which electrode does the sodium form at? Why?

c) What is formed at the other electrode?

7 a) Which of these metals cannot be extracted by reduction with carbon?

potassium
copper
zinc
magnesium
lead
aluminium

How did you decide on your answer?

b) What must we do to extract the metals in part a)?

c) Why are these metals expensive?

8 Which of the statements below are correct?

a) The anodes are made of copper when we extract aluminium.

b) The anodes are positively charged.

c) The aluminium forms at the anode.

d) The aluminium forms at the cathode.

e) When we purify copper the cathode gets heavier in the process.

CHAPTER 5 Metals, Acids and Salts

5a Neutralisation

Acids all around us

What do you think of when you hear the word acid?

a) Talk to your partner about all the things you know or feel about acids.
Draw a mind map of your ideas.

acids

Not all acids are dangerous liquids.
You probably like a little acid on your fish and chips!
Vinegar has the sharp, sour taste of acids.
It contains ethanoic acid.

Vinegar contains ethanoic acid (a weak acid).

b) List 4 things that have a sharp taste.

Does your list include oranges or lemons?
These contain citric acid.
Fizzy drinks are also acidic.

Look at the label from a Coca-Cola bottle:

c) Name an acid in the cola.

Acids in the lab

Have you used any acids in your Science lessons?
Can you remember their names?
There are 3 common acids used in schools.
These are:

hydrochloric acid	HCl
sulphuric acid	H_2SO_4
nitric acid	HNO_3

d) Find out which one of these strong acids is in your stomach.

e) What job does this acid do in your stomach?

Neutralising an acid

Have you ever had indigestion?
That 'burning' feeling comes from too much acid in your stomach.
You can cure the pain quickly by taking a tablet.
The tablet contains an **alkali** (or base) that gets rid of the acid.

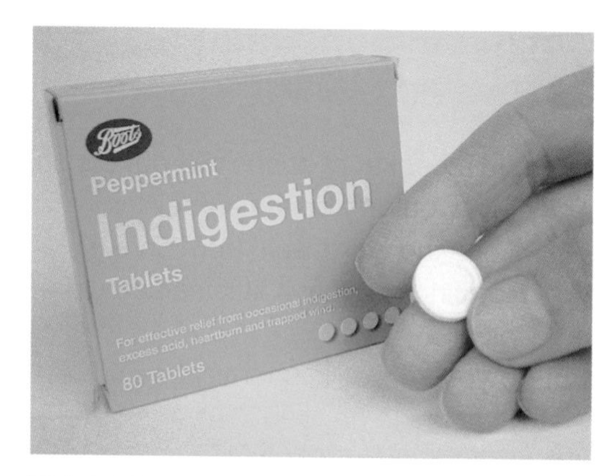

This tablet will neutralise excess acid in your stomach.

f) Name a few types of indigestion tablets.

Acids and alkalis are chemical opposites.
They react together and 'cancel each other out'.

When we mix just the right amount of acid and alkali together, we get a **neutral** solution.

The reaction between an acid and an alkali is called **neutralisation**.

The pH scale

Have you used **universal indicator** before?
It is the indicator that can change to many different colours.
We use it to measure the pH of a solution.
The pH number tells us whether a solution is:

- acidic
- alkaline
- neutral

. . . but also tells us ***how acidic or alkaline*** a solution is.
Look at the pH scale below:

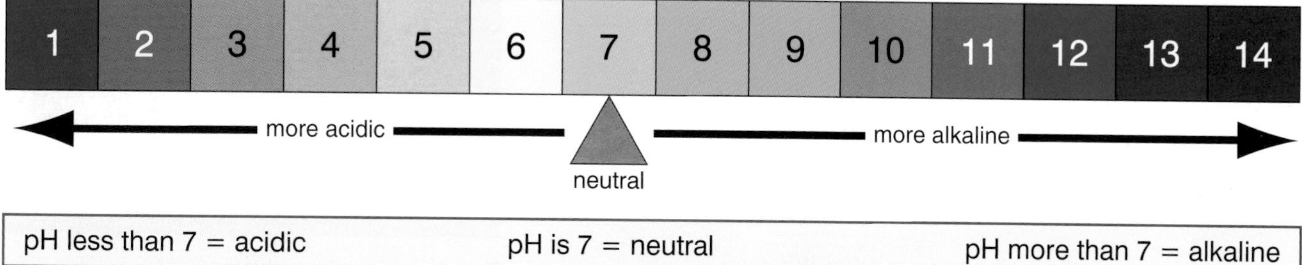

pH less than 7 = acidic	pH is 7 = neutral	pH more than 7 = alkaline

g) Find out 3 ways to measure the pH of a solution.

pH 7,
What colour is seen?
The solution is neutral,
Its colour is green.

Remind yourself!

1 Copy and complete:

Acids and ...... are chemical opposites.
When we add them together, we get a ...... reaction. The ...... number tells us whether a solution is acidic, alkaline or ...... We can use ...... indicator to find this number.

2 a) Put these pH values in order, with the strongest acid first: 5, 8, 2, 12, 7

b) Which one is neutral?

3 Design an advert for a magazine or the TV to sell a new brand of stomach tablet.

5b Salts

Imagine a packet of crisps without the salt.
They just wouldn't be the same!
We all use the word 'salt' from an early age.
We soon find out that a little salt can make some food taste nicer.
But do you know the chemical name of the salt that we sprinkle on food?
It is **sodium chloride**. Its chemical formula is **NaCl** *(nackle!)*.

a) Name a food that tastes better with a little salt.
b) Name a food that adding salt to would spoil.
c) Find out why doctors are worried if we eat too much salt.

But if you ask a chemist what they mean by the word 'salt', you would be surprised at how many salts they could name.
Sodium chloride is just one of a whole group of ***metal compounds called salts***. We can make these salts by neutralising acids.

A **salt** is a metal compound made from an acid.

Crystals of sodium chloride.

d) Why do you think that a chemist might call sodium chloride 'common salt' or 'table salt' rather than just 'salt'?

The salt you make in a neutralisation reaction depends on:

- the acid you use, and
- the metal involved in the reaction.

The metal might be part of an alkali you add to the acid.
Look at the reaction below:

Sodium hydroxide + hydrochloric acid → sodium chloride + water
a salt

e) What is the acid used in the reaction above?
f) Sodium hydroxide is an alkali.
Which metal does it contain?
g) Name the salt made in the reaction.

In general we can say that:

acid + alkali → a salt + water

Naming salts

Look at another reaction of hydrochloric acid:

Magnesium hydroxide + hydrochloric acid → magnesium chloride + water

This is the reaction used when some indigestion tablets work.

h) What is the name of the salt made in the reaction above?

i) What do the names of salts made from hydrochloric acid have in common?

On page 50 we saw the common acids you might find in the lab.

j) Name the 3 common acids found in a chemistry lab.

All acids form salts. The acid gives a salt its 'surname' (the second half of its name).

A salt is a metal plus the 'back-end' of an acid.

Hydrochloric acid makes salts called **chlorides**.
Nitric acid makes salts called **nitrates**.
Sulphuric acid makes salts called **sulphates**.

A salt gets its 'first name' from the metal involved.
So examples of salts are:
lead chloride
zinc nitrate
iron sulphate.

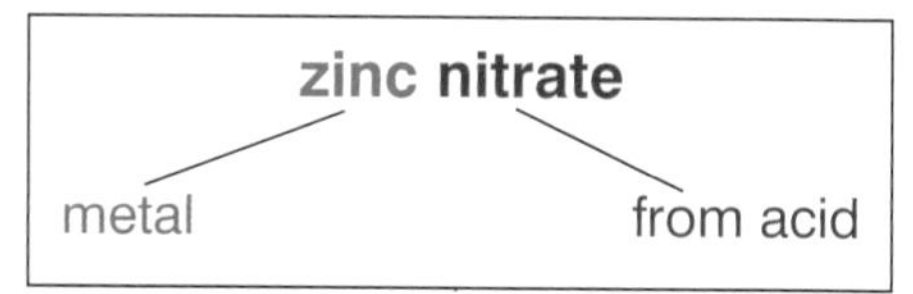

Some metals react with acids.
As they react, hydrogen gas is given off.
A salt is also made.
For example:

zinc + sulphuric acid → zinc sulphate + hydrogen

k) Name the salt made in the reaction above.

Remind yourself!

1 Copy and complete:

Salts are made when an ...... gets ...... by an alkali.
Hydrochloric acid makes salts called ......
Nitrates are made from ...... acid, and ...... are made from sulphuric acid.
The first part of a salt's name comes from a ...... and the second part from an ......

2 Name the salts made in these reactions:

a) zinc + hydrochloric acid

b) magnesium + sulphuric acid

c) iron + nitric acid

3 Explain why you can't make salts of copper or silver by adding the metals to an acid. (Hint: look back to page 28.)

5c Bases and alkalis

Do you know what the opposite of an acid is?
You might say 'an alkali' and you wouldn't be wrong.
But a better answer would be 'a **base**'.
We have already seen how acids are neutralised by alkalis:

an acid + an alkali → a salt + water

We can also say that:

an acid + a base → a salt + water

Lime is a base that we spread onto acidic soil to neutralise it.

A base is the name for any ***compound that neutralises an acid***.
We can think of an alkali as a base that dissolves in water.
Once dissolved, the solution it makes is alkaline.

Alkaline solutions always contain **hydroxide ions** dissolved in the water.
Their formula is **OH^- (aq)**.

Look at the diagram below:

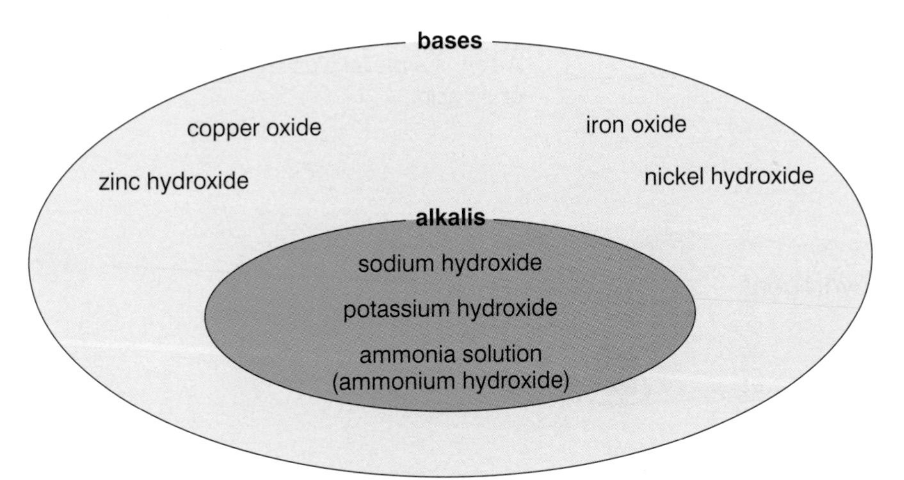

Alkalis in the lab
Here are three alkaline solutions you may come across at school:

- Sodium hydroxide solution
- Potassium hydroxide solution
- Calcium hydroxide solution (called limewater)

a) Name two substances that form alkaline solutions.
b) Name a substance that will neutralise an acid, but does not dissolve in water.
c) What do we call the type of substance named in part b)?
d) What do we get when a base reacts with an acid?
e) Give one use that farmers make of bases.
f) Give a use of a strong alkali.

Strong alkali is used to remove grease from ovens. Why is the man wearing gloves?

Ammonia

Ammonia is a gas.
Its chemical formula is $\mathbf{NH_3}$.
It is the ***only common alkaline gas***.
So we can use this to test for the ammonia gas:

Ammonia gas turns damp red litmus paper blue.

Ammonia is an alkaline gas.

When it dissolves in water it makes a weak solution of ammonium hydroxide. We saw on the previous page that dissolved hydroxide ions make solutions alkaline.

g) How can we test for ammonia gas?
h) Why does the indicator paper have to be damp?

A solution of ammonia (ammonium hydroxide) is an alkali. So it will react with an acid. The salt formed is called an **ammonium salt**.

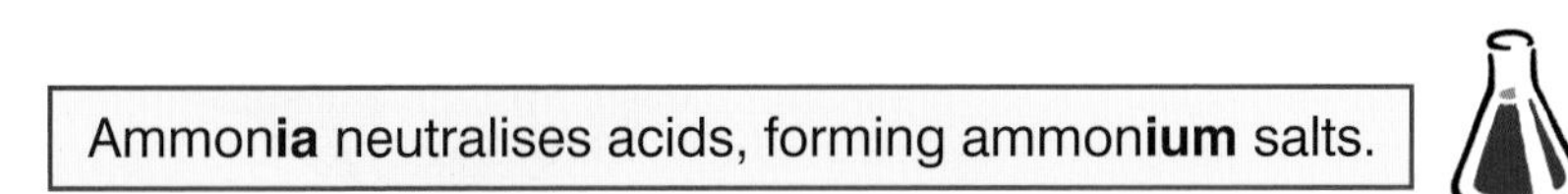
Ammon**ia** neutralises acids, forming ammon**ium** salts.

Ammonia gas and hydrogen chloride gas make the salt ammonium chloride. It is formed as tiny particles in a white 'smoke'.

For example:

ammonium hydroxide + hydrochloric acid → ammonium chloride + water
an ammonium salt

i) Name the salt formed when ammonia solution is added to nitric acid.
j) How could you test if an unknown gas was hydrogen chloride (HCl)?

Remind yourself!

1 Copy and complete:

Bases are the ...... of acids. They react together in a ...... reaction:

acid + base → ...... + ...

Alkalis are bases that are ...... in water. In solution they form ...... ions whose formula is ......
Ammonia gas dissolves in water to form an ...... solution. If an acid is added, we get an ...... salt formed.

2 a) Give the word equation for the reaction between zinc oxide (a base) and sulphuric acid.

b) Write another equation for zinc hydroxide (also a base) plus nitric acid.

3 Name the salt formed in these reactions:

a) nickel oxide + hydrochloric acid

b) copper hydroxide + nitric acid

c) ammonia solution + sulphuric acid

5d Making salts

Salts have many uses. This farmer is treating his vines with copper sulphate to kill pests. Why must grapes be washed well before we eat them?

You've probably used copper sulphate before in chemistry.
It is the compound made up of blue crystals.
Wine-makers sometimes use it to kill pests on their vines.
We can use anhydrous copper sulphate to test for water.
(It is a white powder that turns to blue copper sulphate when you add water.)

a) Is copper sulphate an acid, a base or a salt?
b) What colour is copper sulphate solution?

On pages 52 and 53 we saw how we can make salts from acids.

c) Which acid would you use to make copper sulphate?

Crystals of copper sulphate.

Copper is one of a 'family' of metals called the **transition metals**. (See page 22.) Other examples include iron and nickel.
Their oxides and hydroxides ***don't dissolve in water***, so they can't be alkalis. But they do neutralise acids, and are called **bases**. Look back to page 54:

d) Name a compound of iron that is a base.

We can get the copper in copper sulphate from a base.
We could use the base ***copper oxide*** or ***copper hydroxide***.
The sulphate will come from ***sulphuric acid***.
The salt, ***copper sulphate***, will dissolve in the water present.

e) Finish off this equation:
copper hydroxide + sulphuric acid → ...... + ...

Copper oxide is a base. It neutralises acids, but is insoluble in water.

If you tried this experiment how would you know when all the acid has been used up? How could you tell that it has been neutralised?
You might think that adding an indicator would be a good idea.
But there is no need to. Remember that the base does not dissolve in water. So when all the acid has reacted you can see solid bits of the base in the solution.

f) The base, copper oxide, is a black powder.
What would you see in a beaker when you add more than enough copper oxide to neutralise sulphuric acid solution?

Making copper sulphate in the lab

You can make copper sulphate as shown below:

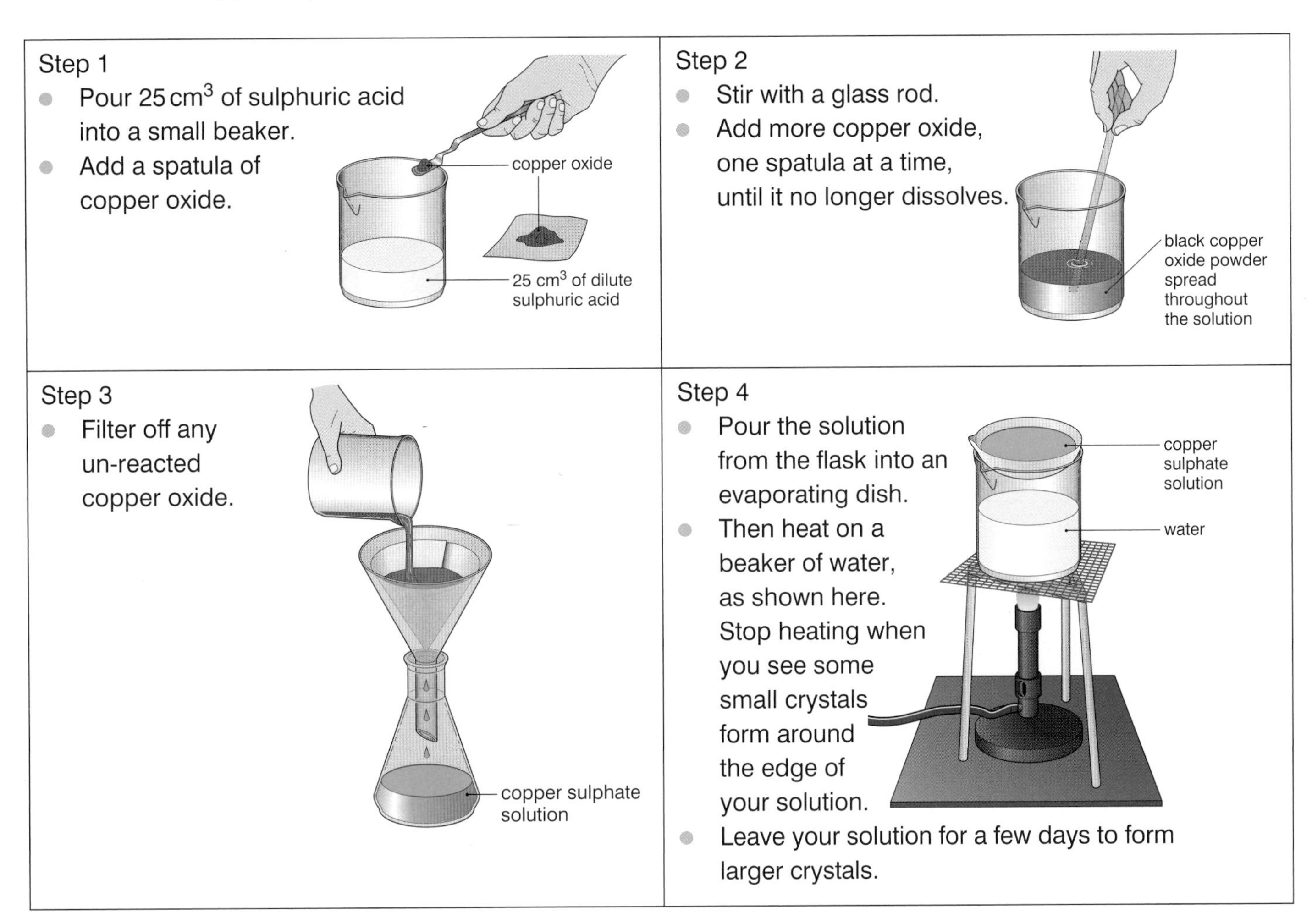

The equation for the reaction above is:

copper oxide + sulphuric acid $\rightarrow$ copper sulphate + water

g) How do you get rid of the excess unreacted base (copper oxide) from the reaction mixture?

Remind yourself!

1 Copy and complete:

Copper oxide and copper ...... are examples of ......
They can be used to n...... acids and make s....... In the reaction, we can see when the acid has been used up because the powder does not ...... in water. You can remove the excess copper oxide by ...... the reaction mixture.

2 Copy and complete these equations:

a) zinc oxide + ... $\rightarrow$ zinc sulphate + ...

b) nickel hydroxide + nitric acid $\rightarrow$ ...... + ...

c) ...... + hydrochloric acid $\rightarrow$ iron ...... + water

3 You are given nickel oxide and dilute hydrochloric acid. Draw a 'step-by-step' method for making nickel chloride.

5e Carbonates with acid

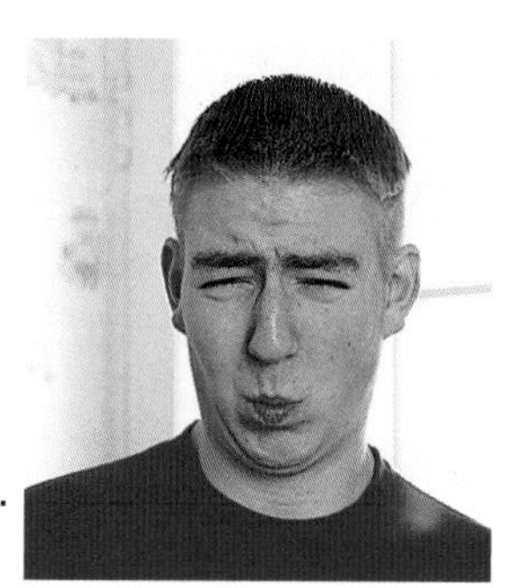

Acids can be neutralised by metal compounds called **carbonates** and **hydrogencarbonates**.

Have you ever tried those sweets or bubble-gum that foam and fizz in your mouth? They contain a weak acid and sodium hydrogencarbonate. When the sweet gets wet as you suck it, some of the acid molecules break down. Like all other acids in water, it makes **hydrogen ions, $H^+(aq)$**. The acid can then react with the hydrogencarbonate. The fizzing is caused by the carbon dioxide gas given off.

This reaction is also used in baking powder. It makes cakes rise.

In general the reactions are:

acid + a carbonate → a salt + water + **carbon dioxide**

and

acid + a hydrogencarbonate → a salt + water + **carbon dioxide**

a) Which ions do we find in all acidic solutions?
b) Which gas is given off when a carbonate reacts with an acid?
c) Explain how baking powder makes cake-mix rise.

Baking powder helps a cake rise to the occasion!

The carbonates of the alkali metals dissolve in water. But the carbonates of the transition metals are insoluble. So we can use a method like the one shown on the previous page to make a transition metal salt. You will see solid carbonate powder left when the acid has been neutralised. The fizzing will also stop. Look at the experiment below:

- Stir with a glass rod.
- Add more copper carbonate, one spatula at a time, until it no longer fizzes.

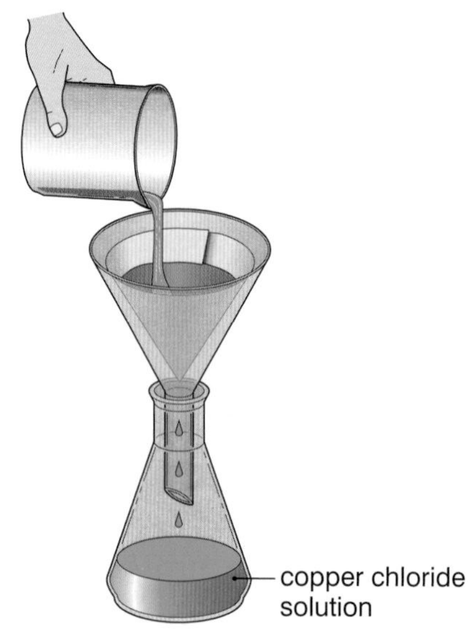

- Filter off any un-reacted copper carbonate.

d) Name the salt made in the experiment.

e) How would you get crystals of the salt?

f) Why is the reaction mixture filtered?

g) How would you know when the reaction between an acid and sodium carbonate had finished?

Useful neutralisation

We have already seen how indigestion tablets work on page 51.
They neutralise excess acid in your stomach.
There are also other useful neutralisation reactions.

Look at the photos below:

Sometimes the soil might be too acidic for a particular crop.

Coal-fired power stations give off sulphur dioxide gas. This causes acid rain.

Farmers neutralise acidic soil using powdered limestone (calcium carbonate) or slaked lime (calcium hydroxide).

We can help reduce acid rain by stopping acidic gases escaping from factories. The gases given out are passed through 'scrubbers'. These contain calcium carbonate mixed with water.

h) Why do farmers neutralise acidic soil?

i) Sulphur dioxide is the main cause of acid rain. What is its chemical formula?

Remind yourself!

1 Copy and complete:

Acids react with hydrogencarbonates and ......
giving off ...... dioxide gas.
You can use the reaction to make s......
For example, ... chloride can be made from copper carbonate and ...... acid.

All acids produce ...... ions in water.

2 Complete these word equations:

a) nitric acid + ... carbonate →
calcium nitrate + ... + water

b) sulphuric acid + ... carbonate →
zinc ...... + ... + ...

3 Use the information in this chapter and other sources to make a poster called 'Useful reactions of acids'.

Summary

We can **neutralise** an acid by reacting it with a base.
Alkalis are bases that can dissolve in water.
Acids form **hydrogen ions**, $H^+(aq)$, and alkalis form **hydroxide ions**, $OH^-(aq)$.
The general equation for a neutralisation reaction is:

acid + a base (or alkali) $\rightarrow$ a salt + water

The salt we make depends on:

1. the acid you use, and
2. the metal in the base or alkali.

The salt gets the first part of its name from the metal.
The last part of its name comes from the acid.
Hydrochloric acid makes salts called **chlorides**.
Nitric acid makes salts called **nitrates**.
Sulphuric acid makes salts called **sulphates**.

We can make crystals of metal salts from acids.
With an insoluble base, such as a transition metal hydroxide (or oxide), we can filter off the excess base after the acid has been neutralised.
Then we evaporate off some of the water from the salt solution and leave it long enough for the crystals to form.

Example of preparing a salt from a carbonate

1.
- Pour 25 cm^3 of hydrochloric acid into a small beaker.
- Add a spatula of copper carbonate.

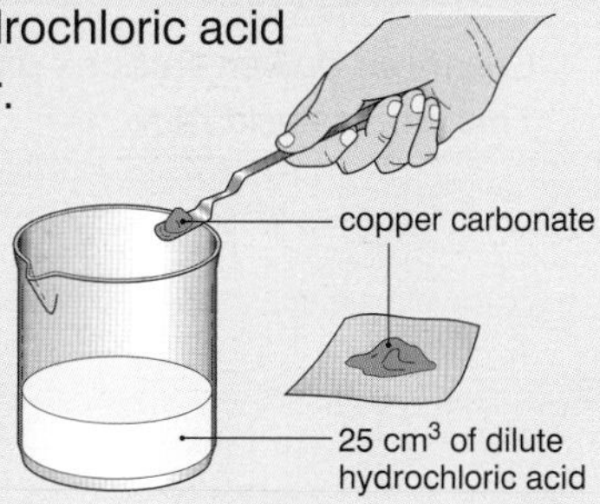

copper carbonate
25 cm^3 of dilute hydrochloric acid

2.
- Stir with a glass rod.
- Add more copper carbonate, one spatula at a time, until it no longer fizzes.

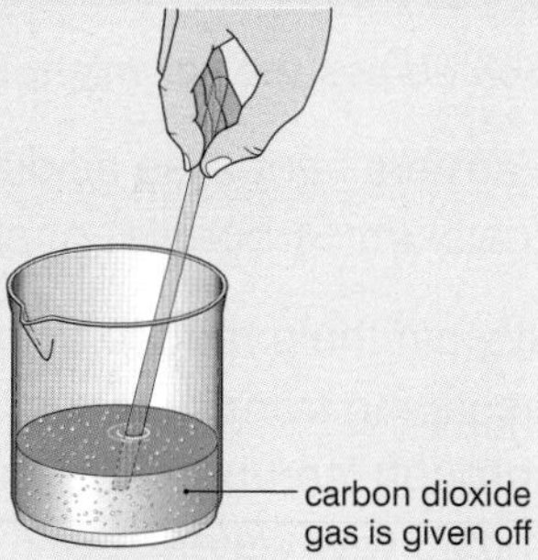

carbon dioxide gas is given off

3.
- Filter off any un-reacted copper carbonate.

copper chloride solution

4.
- Pour the solution from the flask into an evaporating dish.
- Then heat on a beaker of water, as shown here. Stop heating when you see some small crystals form around the edge of your solution.
- Leave your solution for a few days to form crystals.

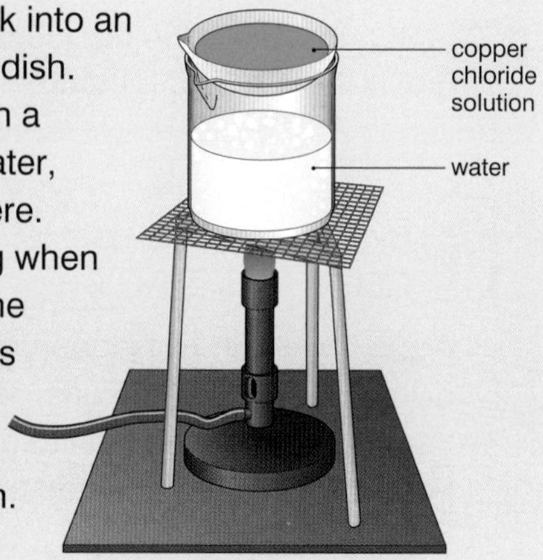

copper chloride solution
water

With an alkali we have to use an indicator to see when the acid has been neutralised.

Questions

1 Copy and complete:

Acids can be …… by bases. Bases that are soluble in …… are called …… These react with acids as shown below:

acid + …… → …… + water

Different acids make different salts. For example, …… acid makes chlorides, nitric acid makes ……, and sulphuric acid makes …… So sodium …… can be made from nitric acid and magnesium …… from sulphuric acid.

All acids produce …… ions in water. Their formula is ……(aq). Alkalis form ……(aq) ions called …… ions.

2 Gemma tested some solutions with universal indicator.

a) The first solution turned purple with the indicator. Was the solution acidic, alkaline or neutral?

b) The next solution was green with the indicator. Was it acidic, alkaline or neutral?

c) Then she tested another 4 liquids and wrote down their pH values:

1, 5, 7, 14

but forgot to write the names of the liquids. Help her to fill in the table below:

Liquid tested	pH value
sulphuric acid	
distilled water	
sodium hydroxide	
vinegar	

d) What colour would each of the solutions turn when universal indicator is added?

e) Name another way that we can test the pH of a solution.

f) Which method of measuring pH is better?

Explain your answer.

3 Joanne and Sakib want to make some crystals of copper sulphate.
They have dilute sulphuric acid and black copper oxide powder.

a) Write a word equation for the reaction.

b) How could they tell when all the sulphuric acid has been used up in the reaction?

c) What type of substance is the black copper oxide – an acid, a base or an alkali?

d) Write a method that someone in another class could follow to make copper sulphate crystals. You can draw diagrams to help.

4 The table below shows the pH of the soil that some plants grow well in:

Plant	pH
apple	5.0–6.5
potato	4.5–6.0
blackcurrant	6.0–8.0
mint	7.0–8.0
onion	6.0–7.0
strawberry	5.0–7.0

a) Which plant can grow in the most acidic soil?

b) Which plants can grow in alkaline soil?

c) How can you neutralise an acidic soil?

5 Complete these word equations:

a) sodium hydroxide + hydrochloric acid → …… …… + ……

b) zinc hydroxide + sulphuric acid → …… …… + ……

c) copper carbonate + nitric acid → …… …… + …… + …… ……

6 a) Ammonia gas dissolves in water.

Is the solution formed acidic, alkaline or neutral?

b) Name the salt made if ammonia solution reacts with nitric acid.

Further questions on Metals

Groups of metals

1 **(a)** The table shows some properties of metals **and** non-metals.

Four of these are properties of metals. One of these properties has been ticked.

Which are the **other three** properties of metals.

PROPERTY	
mostly have a high melting point	
are gases at room temperature	
are good conductors of electricity	✓
are poor conductors of heat	
are shiny when freshly cut	
are mostly brittle when solid	
mostly have a low boiling point	
can be bent into shape when solid	

(3)

(b) Metals are used to make central heating radiators.

Explain, as fully as you can, why metals are good materials for making radiators. (3)

(AQA 2001)

2 The metals aluminium and copper have many uses. Give **two** properties of these metals which makes them best suited for the uses shown.

(a) The use of aluminium for making cooking pans. (2)

(b) The use of copper for making electrical wiring. (2)

(AQA 1999)

3 The table below shows some properties of four elements.

Element	Melting Point (°C)	Boiling Point (°C)	Electrical Conductivity	Density (g/cm^3)
copper	1083	2600	conductor	8.9
iodine	114	184	non-conductor	4.9
iron	1540	3000	conductor	7.9
sulphur	119	445	non-conductor	2.1

Use the table above to answer the following questions.

(i) Give the name of **two** elements which are metals. (1)

(ii) Give **two** reasons for your choice of elements in part (i) (2)

(WJEC)

4 The list below shows some Group I elements.

lithium

sodium

potassium

Select words from the box below to complete the following sentences about Group I elements.

decreases	hydrogen	increases
oil	oxygen	

(i) The reactivity of Group I elements ...... from the top to the bottom of the group. (1)

(ii) Group I elements are stored in ...... to prevent them reacting with air. (1)

(iii) Potassium reacts with ...... in the air forming potassium oxide. (1)

(iv) Sodium reacts violently with water forming the colourless gas .... (1)

(WJEC)

5 **(a)** Describe the reaction of sodium with cold water. Include **observations** and **name the products** of the reaction. (4)

(b) Lithium reacts in a similar way to sodium. Complete the **word** equation for the reaction between lithium and water.

lithium + water → ...... + ... (2)

(c) Lithium and sodium are usually stored in liquid paraffin (oil).

Give **one** reason for this method of storage. (1)

(d) Sodium burns vigorously in oxygen to form its oxide.

Copy and balance the **symbol** equation for the reaction.

...... Na + ... → ...... Na_2O (2)

(e) Would you expect potassium to react **more, the same or less** vigorously than sodium in its reaction with oxygen? Give your reasoning. (2)

(WJEC)

Further questions on Metals

Extraction of metals

6 By observing the reactions of metals with water and dilute sulphuric acid it is possible to put metals in order of their reactivity.

(a) **A**, **B**, **C** and **D** represent four metals.

Metal	Reaction with water	Reaction with dilute sulphuric acid
A	No reaction	Reacts slowly at first
B	No reaction	No reaction
C	Little or no reaction	Reacts quickly
D	Vigorous reaction	Violent – dangerous reaction

(i) Put metals **A**, **B**, **C** and **D** in order of their reactivity (most reactive first). (2)

(ii) The metals used were copper, magnesium, sodium and zinc. Use the information in the table to identify which of these metals was **A**, **B**, **C** and **D**. (2)

(AQA SEG 2000)

7 The reaction between aluminium and iron oxide is used to join lengths of railway track. It is called the thermit reaction.

$Fe_2O_3(s) + 2Al(s) \rightarrow Al_2O_3(s) + 2Fe(l)$

(i) Why does aluminium react with iron oxide? (1)

(ii) What does the (l) after Fe in the chemical equation mean? (1)

(iii) Suggest why the thermit reaction can be used to join lengths of railway track. (2)

(AQA SEG 1999)

8 **(a)** Iron is extracted from its ore in a blast furnace. Choose words from the box to label the diagram of the blast furnace.

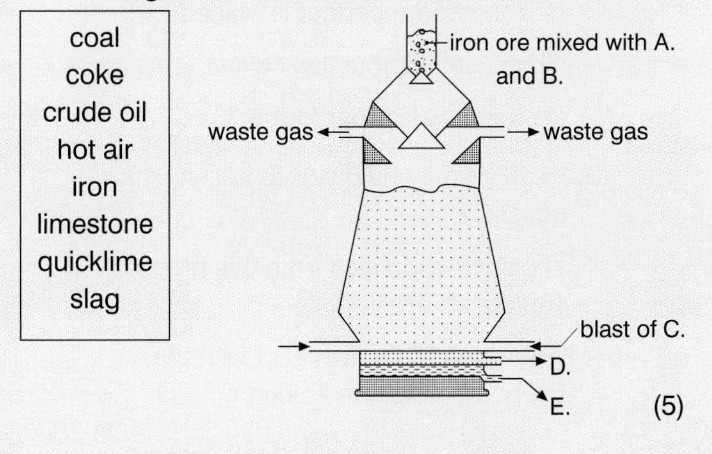

(5)

(b) The word equations for some of the reactions that take place in the blast furnace are given below.

A carbon + oxygen → carbon dioxide

B iron oxide + carbon monoxide → iron + carbon dioxide

C calcium carbonate → calcium oxide + carbon dioxide

D calcium oxide + silicon oxide → calcium silicate

In which of the reactions, **A**, **B**, **C** or **D**, is the first named substance reduced? (1)

(AQA SEG 2000)

9 **(a)** Use words from the box to complete the paragraph.

carbon	coke	iron	limestone
oxygen	slag	steel	titanium

Production of Steel

The metal produced in the blast furnace is

(i) .... It contains about 4% carbon. Some of this carbon must be removed to produce

(ii) .... This is done by blowing

(iii) ...... through the molten metal.
To produce alloys, other metals such as

(iv) ...... may be added. (4)

(b) Complete each of the following sentences.

(i) **Steel** is better than **cast iron** for making car bodies because steel ...... (1)

(ii) **Stainless steel** is better than **mild steel** for making saucepans because stainless steel ...... (1)

(iii) **Aluminium alloy** is better than **stainless steel** for making aeroplanes because aluminium alloy ...... (1)

(EDEXCEL 1999)

Metals and electrolysis

10 The diagram shows a method of producing aluminium.

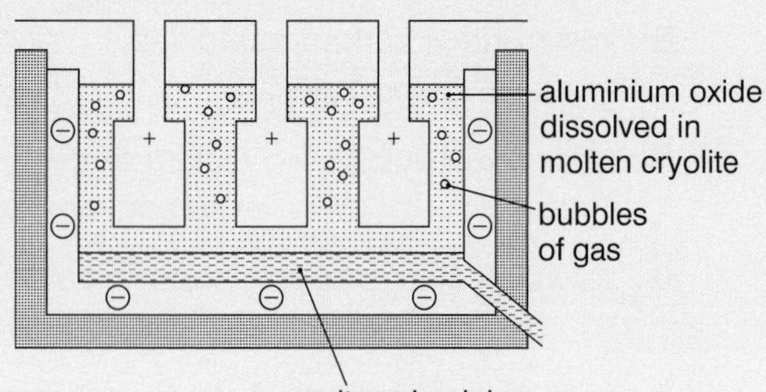

(a) The positive and negative electrodes are made of the same element.
Name this element from the box below: (1)

aluminium	carbon	iron	zinc

(b) Aluminium oxide is dissolved in molten cryolite.
Choose the reason why cryolite is used from the list below: (1)

It acts as a catalyst.

It makes purer aluminium.

It lowers the melting point of the aluminium oxide.

It increases the life of the electrodes.

(c) Suggest **one** reason why aluminium is expensive to produce. (1)

(d) Complete the sentence by choosing from the list.

burn away quickly

need regular cleaning

stop conducting electricity

increase in size

The positive electrodes have to be replaced frequently because they . . . (1)

(e) In the reaction, the ions O^{2-} and Al^{3+} are attracted to the electrodes.

Copy the following diagram and draw arrows to show the directions in which the ions move.

(1)

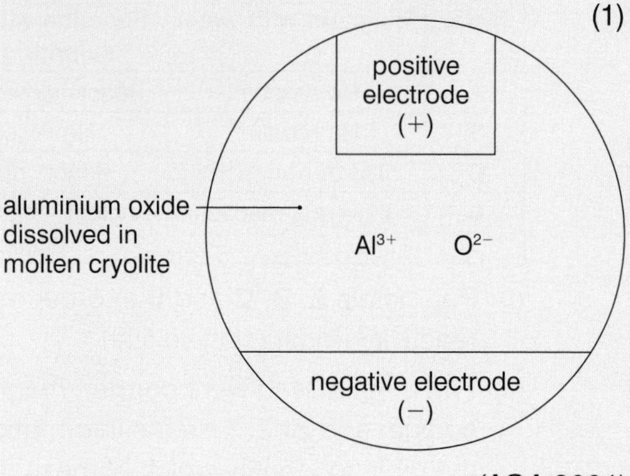

(AQA 2001)

11 **(a)** The diagram shows a method for obtaining pure copper from impure copper.

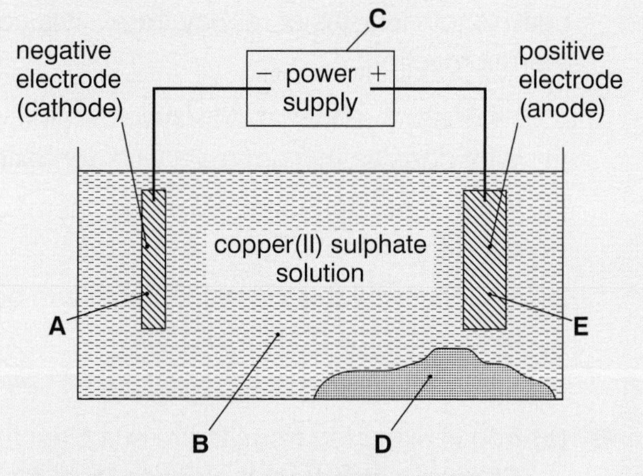

State the place, **A**, **B**, **C**, **D** or **E**, where:

(i) the impure copper is placed

(ii) the solid impurities collect

(iii) the pure copper forms (3)

(b) Page 28 may help you to answer this question.

The solid impurities from this process contain silver.

Why does silver **not** react with the copper(II) sulphate solution? (1)

(AQA 2000)

Metals, acids and salts

12 **(a)** The chart shows the pH ranges at which some crops grow best.

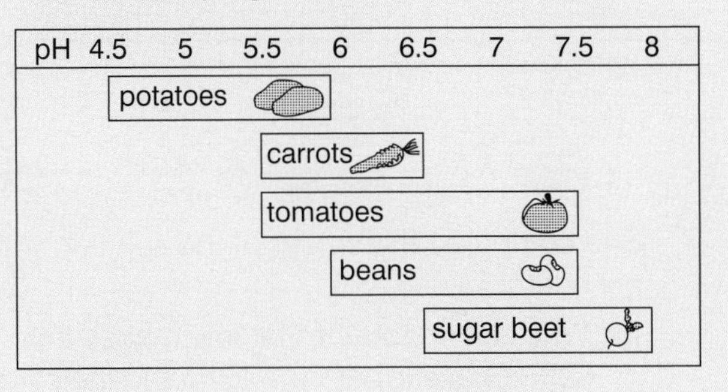

(i) What does pH measure? (1)

(ii) What is the pH of a neutral solution? (1)

(iii) What colour is universal indicator in a neutral solution? (1)

(iv) Which **two** crops grow best only in acidic soils? (2)

(b) Describe how a farmer could test the pH of a sample of soil using the substances and apparatus below:

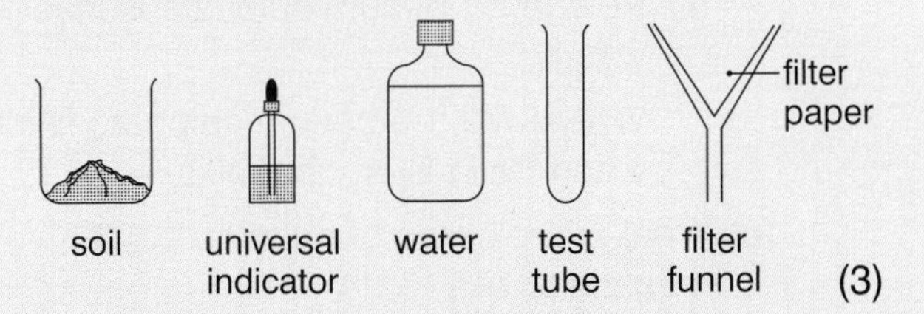

(3)

(c) The farmer tested the pH of the soil from a field in which sugar beet was to be grown.

Explain why the farmer then wanted to spread lime on the field. (2)

(AQA SEG 1998)

13 This question is about acids and bases.

(a) The table shows some substances and their pH values.

Substance	pH value
grapefruit juice	3.1
blood	7.4
sea water	8.5
beer	5.2
oven cleaner	9.8

Choose substances from the list to answer these questions.

(i) Which substance is the most acidic? (1)

(ii) Which substance has a pH value closest to neutral? (1)

(iii) Which substance is the strongest alkali? (1)

(b) A sherbet sweet contains citric acid, sugar and sodium hydrogencarbonate.

When you put the sweet into your mouth, it mixes with water and fizzes.

Which **two** substances react to produce the fizzing? (1)

(c) When a nettle stings you, it injects you with a small amount of acid.

If you rub the sting with a dock leaf, the sting seems to hurt less.

What sort of substance is in the dock leaf?

Choose from the list below: (1)

acid **alkali** **salt**

(OCR 1999)

14 A student tried to make some magnesium sulphate. Excess magnesium was added to dilute sulphuric acid. During this reaction fizzing was observed due to the production of a gas.

(i) Complete and balance the chemical equation for this reaction.

...... + $H_2SO_4 \rightarrow$ ...... + ... (3)

(ii) At the end of the reaction the solution remaining was filtered. Why was the solution filtered? (1)

(iii) The filtered solution was left in a warm place.

Explain why the filtered solution was left in a warm place. (2)

(AQA SEG 2000)

15 Dilute hydrochloric acid was added slowly to dilute sodium hydroxide solution in a beaker. The graph below shows how the pH of the solution in the beaker changed as the acid was added.

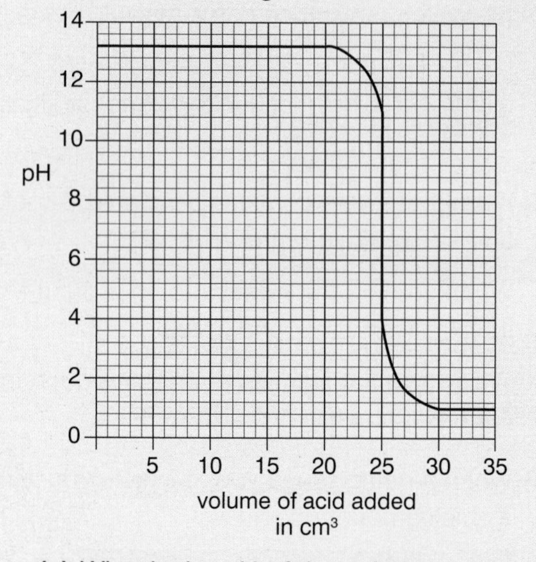

(a) What is the pH of the solution in the beaker when 30 cm^3 of dilute hydrochloric acid has been added? (1)

(b) The dilute sodium hydroxide solution in the beaker contained Universal indicator.

What colour was the solution in the beaker when the following volumes of dilute hydrochloric acid had been added?

(i) 30.0 cm^3 (1)

(ii) 10.0 cm^3 (1)

(c) (i) What is the pH of a neutral solution? (1)

(ii) What volume of dilute hydrochloric acid was added to neutralise the sodium hydroxide solution in the beaker? (1)

(iii) The neutral solution was evaporated to dryness to leave a solid salt.
What is the name of the salt which is formed? (1)

(iv) Describe what the salt looks like. (1)

(v) Complete the word equation for the reaction of sodium hydroxide with hydrochloric acid.

sodium hydroxide + hydrochloric acid → ...... + ...

(1)

(EDEXCEL 1999)

16 Bella wants to test the pH of soil in her garden. She takes a sample of soil from her garden.

The pictures show what she does.

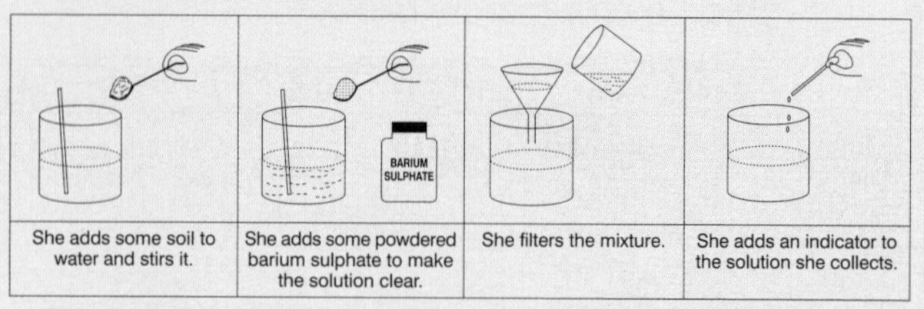

She compares the colour of the solution with a pH chart.

(a) Why is it important that she makes the solution clear before adding the indicator? (1)

(b) The diagram shows some of Bella's apparatus.

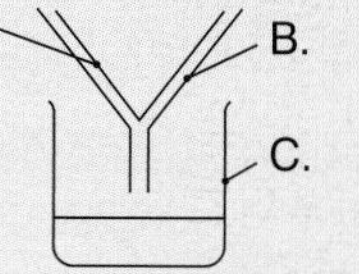

(i) What are the **three** labels (A, B and C) on this apparatus.

Choose from this list.

beaker filter paper funnel
stirring rod test tube thermometer (3)

(ii) Powdered barium sulphate is insoluble in water.

Where will the barium sulphate finish up when Bella filters the mixture. (1)

(c) What is the name of an indicator for measuring pH.

Choose from the list below:

litmus starch universal (1)

(d) She tests two other samples of soil from different places in her garden.

Bella labels the three samples of soil **A**, **B** and **C**.

Why does she test three samples of soil? (1)

(e) The chart shows her results.

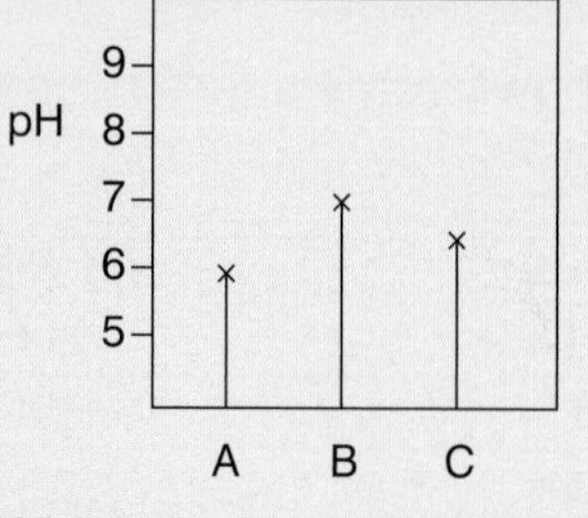

Which sample of soil was exactly neutral (A, B or C)? (1)

(OCR Nuffield 1999)

Section Two

The Earth and its resources

In this section you will find out more about our planet and the important materials it provides for us. You will look at the products we get from limestone and crude oil. Then you can study the Earth itself: what it is made from and its surrounding atmosphere.

CHAPTER 6

6a Limestone – a useful rock

It's hard for us to imagine the time it took for limestone to form millions of years ago. The rock is made mainly from the shells of sea creatures. These were crushed together as layers of sediment built up on ancient sea beds and eventually turned to rock.

There are different types of limestone but all of them contain over 50% **calcium carbonate**.
Its formula is $\mathbf{CaCO_3}$. Chalk is a type of limestone and marble is made from limestone.

Fossils in limestone.

a) How many chemical elements make up calcium carbonate?

b) Name 3 types of rock that contain calcium carbonate.

Limestone is mainly used in the building industry.
If your school is made from bricks, they will be held in place with mortar. This is made with cement which we make from limestone.
Or perhaps your school is made from concrete?
Again, this is made using cement. The steel rods that strengthen reinforced concrete are made using limestone as a raw material.
Even the glass in the windows was made using limestone.

Look at the diagram below:

glass in window

mortar

iron gates

steel rods embedded in concrete floors

Products using limestone.

c) List 4 materials used in the building industry that use limestone as a raw material.

d) What caused the problem that some schools had with the concrete in their buildings? (Hint: think about rusting.)

The limestone for all these uses is quarried out of the ground. Have you ever seen a quarry? They are not a pretty sight! They produce a huge scar on the landscape.

Imagine a limestone company want to build a new quarry in a beautiful piece of countryside.
Look at some of the arguments below:

A limestone quarry.

Remind yourself!

1 Copy and complete:

Most limestone was formed ...... of years ago from the ...... of sea creatures.
Its chemical name is calcium ...... and its formula is ......
We make many useful building materials from limestone including cement, c......, g...... and s......

2 Imagine you are a journalist for the local newspaper living in the village above.
Write an article about the proposal for a new quarry.
You can use the comments above, but feel free to make up your own!

6b Heating limestone

Heating calcium carbonate.

Have you ever heated limestone in the chemistry lab?
Look at the photo opposite:

The orange glow you see was once used to light up the stage in theatres (before we had electricity, of course!). That's where our saying 'To be in the limelight' comes from.

When we heat up the limestone a chemical reaction happens.
The calcium carbonate breaks down.
It forms calcium oxide and carbon dioxide gas.
Look at the equation below:

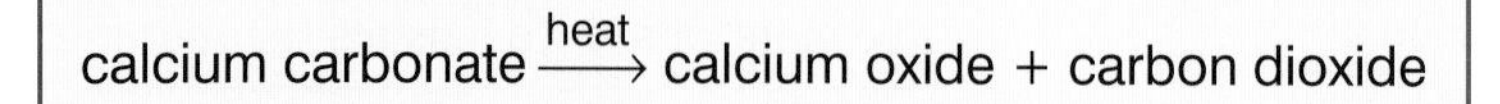

calcium carbonate $\xrightarrow{\text{heat}}$ calcium oxide + carbon dioxide

The balanced equation is:

$$CaCO_3(s) \longrightarrow CaO(s) + CO_2(g)$$

This type of reaction is called **thermal decomposition**.

Thermal decomposition is the breakdown of a substance by heat.

a) Name the products when we heat limestone.

b) Other carbonates are also broken down by heat in the same way.
Predict the word equation when we heat copper carbonate.

c) What do we call this type of reaction?

d) The formula of copper carbonate is $CuCO_3$.
Write a balanced equation for the reaction in question b).

The calcium oxide formed in the breakdown of limestone is called **quicklime**. This name comes from the old English word that meant 'living'. Look at the photo opposite:

When you add a little water to calcium oxide, it expands and starts to crumble up. This movement makes it look like the calcium oxide has a life of its own (or so people thought hundreds of years ago!).
Here is the word equation:

calcium oxide + water → calcium hydroxide
(quicklime) **(slaked lime)**

e) Do you think calcium hydroxide is acidic, alkaline or neutral when it dissolves in water? (See page 54.)

Adding water to calcium oxide.

The Lime Kiln

Of course in industry we don't heat limestone with Bunsen burners! (Although gas is used as the fuel to heat lime kilns.)
Look at the kiln below:

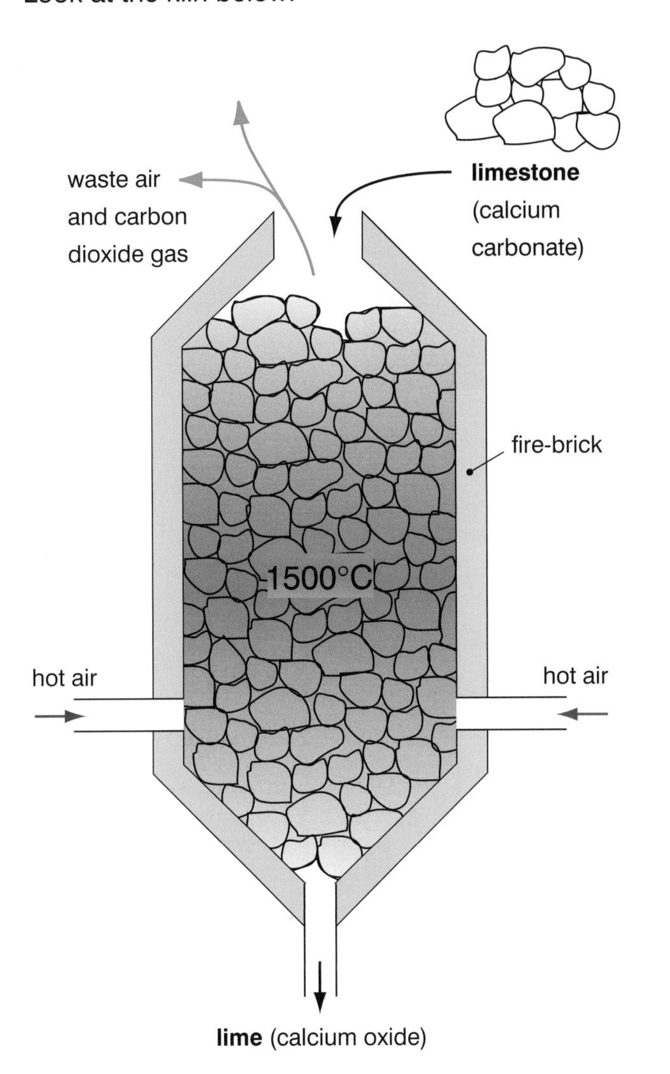

f) Which gas is given off from a lime kiln?

g) How does too much of this gas affect our environment? (See page 89.)

Remind yourself!

1 Copy and complete:

We heat limestone inside a lime ......
The limestone breaks down to form calcium ...... (known as quicklime) and .... gas. This type of reaction is called ...... decomposition.

Adding water to quicklime gives us ...... lime (calcium ......) which is a cheap alkali.

2 Copy and complete:

calcium carbonate
↓ heat
.... + carbon ......
↓ add water
....

6c Cement

Have you ever passed a cement works?
The fine powder that covers the site gives us a clue that limestone is used to make cement.
Cement is a very important material for builders.
They use the grey powder to make their mortar and concrete.

Making cement

We make cement by heating powdered limestone and clay (or shale).
The powder is heated in large kilns that rotate.
Look at the diagram below:

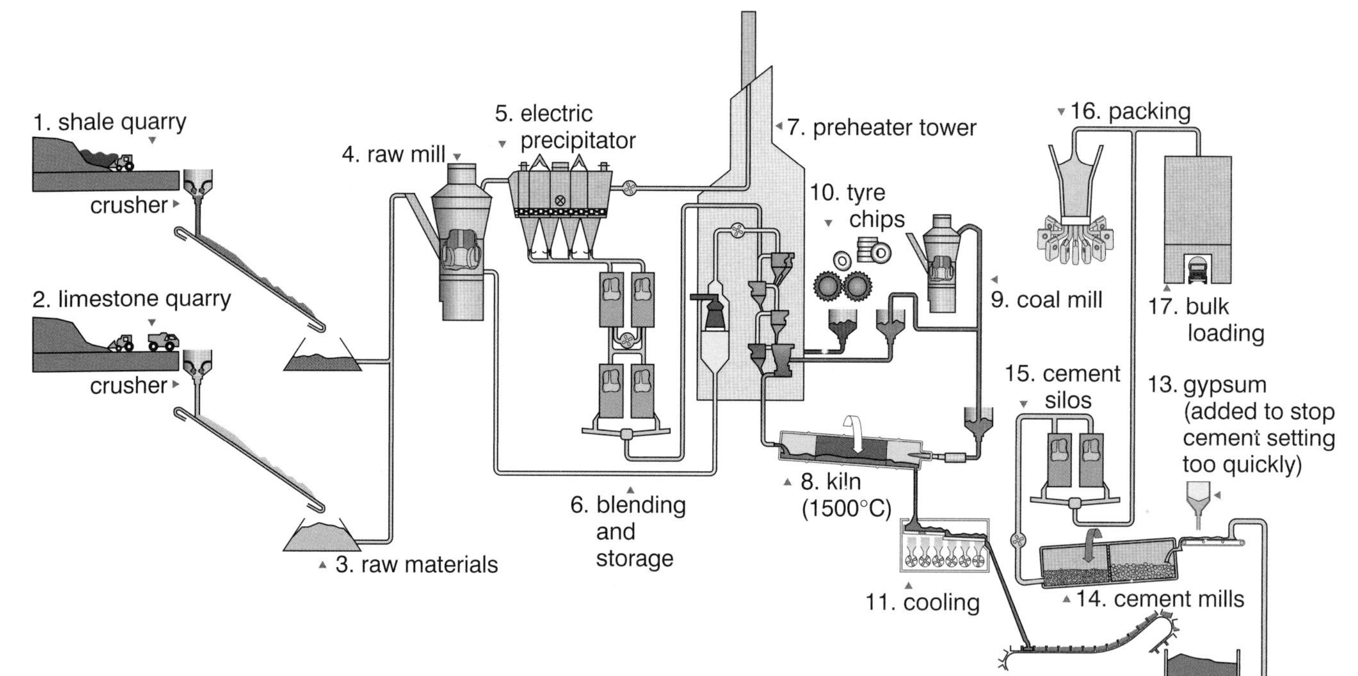

A rotary kiln to heat the powdered limestone and clay (or shale).

a) What happens to the limestone and shale when they arrive at the factory?
b) What is the temperature inside the rotary kiln?
c) How does the powder pass along the kiln?
d) Why is a little gypsum (calcium sulphate) added to cement?

Mixing mortar

Have you ever watched a brick-layer at work?
They set the bricks in place using mortar.
Somewhere near the brick-layer you will see the mortar being mixed.
Mortar is made by mixing cement and sand together.
Then you have to add just the right amount of water – not too much or not too little.
The mortar sets overnight. But it will carry on reacting for several months, getting stronger and stronger.

e) What do we add to cement powder to make mortar?
f) What would happen if the mortar was too runny?
g) What would happen if the mortar was too thick?

Hollywood actors are invited to record their hand prints on the sidewalk once they are famous.

Mixing concrete

The most widely used building material in the world is concrete. You will have seen the specially designed lorries carrying it to building sites.

We make concrete by ***mixing cement with sand and crushed rock (or small stones), then adding water***.
As with mortar, the reaction to set the concrete is very slow and carries on over months.
Look at some uses of concrete below:

h) What is in the mixture that makes concrete that is not in mortar?

Remind yourself!

1 Copy and complete:

Cement is made by heating ...... limestone and ...... (or shale) in large ...... kilns.
The cement is mixed with ...... and crushed ......, followed by water, to make ......

2 a) Design an investigation to find out which mixture makes the best concrete.
Think about how to vary the number of parts of cement, sand and stones, and how you will test the strength of the concrete made.

b) How do builders make concrete stronger?

6d Glass

Think of all the things we use that are made from glass:
Look at the pictures below:

a) Write down two more uses of glass.

We use limestone when making glass.
But did you know that the main material used is sand?
Glass was probably discovered in the sand underneath a fire thousands of years ago.
The first glass object has been dated at about 4500 BC.
We know that the ancient Egyptians used glass containers around 5000 years ago.

b) Work out a rough date when the ancient Egyptians were using glass containers.

Look at the raw materials we use to make glass today:

> Glass is made from:
> - Sand
> - Limestone, and
> - Soda (sodium carbonate)

These are heated up to about 1500°C and the mixture reacts to form molten glass.
As well as these materials, **re-cycled glass** is now making up more and more of the glass making mixture.

Re-cycled glass is becoming more and more important.

Different types of glass

Have you ever seen a car window that has smashed?
It looks a lot different from a broken glass bottle.
That's because the glass is made in different ways.
A windscreen is made like a glass sandwich!
The filling in the sandwich is a thin sheet of plastic.
It is called laminated glass.

c) Why is it important to use laminated glass for a car windscreen?

Glass makers have always experimented to find ways to improve glass.
Look at the table below?

Type of glass	Use
soda-lime	windows
boro-silicate	test-tubes, beakers (heat-proof, chemical resistant)
lead-crystal	wine glasses, bowls, vases
glass fibres	fibre optics, fibreglass
optical glass	lenses in spectacles, cameras, projectors, etc.
glass ceramic	opaque oven-ware

d) Which types of glass are heat-proof?
e) Think of another use for optical glass.
f) Find out a use for glass fibres and one for fibreglass.

We can also make coloured glass.
You just add a small amount of a transition metal compound.
Different metals give different colours.

Remind yourself!

1 Copy and complete:

Glass is made by heating a mixture of ......, ... and ...... (sodium ......) to a temperature of about ......°C.

...... glass is being used more and more in glass making mixtures.

2 a) How can you make a glass that won't shatter when it gets hit?

b) What is this type of glass called?

c) What is this glass used for?

3 Design a leaflet to encourage shoppers in a supermarket to re-cycle glass.

6e Neutralising acidic soil

Have you ever seen fields with the soil covered in white powder (not snow!)?
These fields have been treated with powdered limestone or slaked lime.
They are both used to ***neutralise acidic soil***.

Neutralising acidic soil.

a) Does the lime or limestone added raise or lower the pH of the soil?

b) What are the chemical names for limestone and slaked lime?

c) Look back to the last chapter:
What will be formed when limestone (a carbonate) reacts with an acid?

d) What will be formed when slaked lime (an alkali) reacts with an acid?

We also use powdered limestone in ***lakes affected by acid rain***.
In Norway and Sweden many lakes now are too acidic for fish to live.
Look at the photo below:

This lake has become acidic.
It is being neutralised by calcium carbonate.

Summary

Limestone is made up mainly of **calcium carbonate**.
It is used as a building material itself, but it also makes **cement**.
Powdered limestone is heated in a rotating kiln with clay (or shale) to make the cement.

Cement, sand and bits of rock make **concrete** – the most widely used building material.
This is made by mixing cement, sand and crushed rock, together with water.

When we heat limestone in a lime kiln it makes quicklime.
The reaction is called **thermal decomposition**.
By adding water to quicklime, we get slaked lime – a cheap alkali.
This, or powdered limestone, can be used to neutralise acidic soil.

We make glass by heating sand, limestone and soda (sodium carbonate).

Questions

1 Copy and complete:

Calcium ...... is found naturally in limestone rock. When heated in a lime kiln there is a thermal ...... reaction. We get ...... (calcium oxide) formed and carbon ...... gas is given off.

If a little water is added to calcium oxide we get ...... lime (calcium ......). This can be used to ...... acidic soil.

Powdered limestone is heated with ...... (or shale) to make cement. This can be mixed with sand and ...... rock, followed by water to make ......

Glass is made from ......, ... and ......

2 What is formed when we heat these compounds?

a) calcium carbonate → .... + ...

b) magnesium carbonate → .... + ...

c) nickel carbonate → .... + ...

d) What are these types of reaction called?

e) Draw a diagram to show how you could test the gas given off during one of the reactions above.

3 Here are the results of an experiment to see which mix of mortar is strongest:

Mixture	Height of weight dropped on the mortar before breaking (cm)
A	17
B	25
C	36
D	13
E	10

a) Describe how you think the experiment was carried out.

b) Draw a suitable graph to show these results.

c) Which mixture was best?

d) What would you add to the mortar mix to make concrete?

e) What is set inside the concrete to reinforce it?

f) i) Name something that would be built using reinforced concrete.
ii) Name something else that doesn't need reinforced concrete.

CHAPTER 7

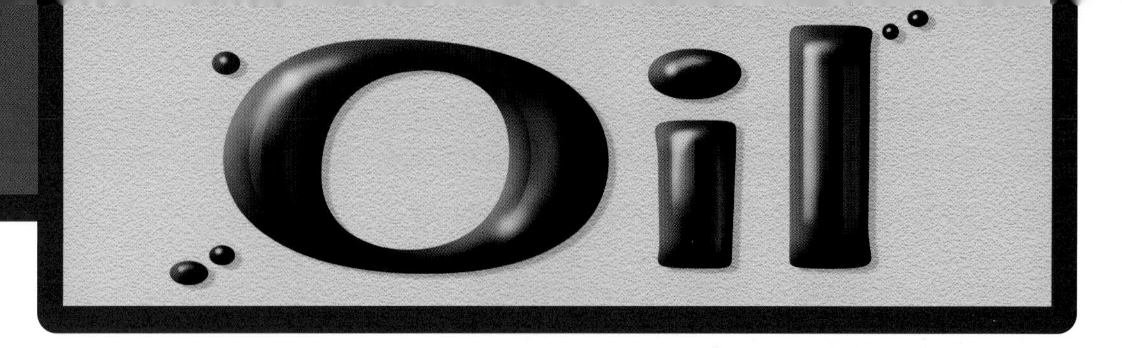

7a The story of oil

Life without oil, and the products we get from it, is hard to imagine. Our lives would be very different without petrol, diesel, plastics and many other useful things we make from crude oil. Look at the photos below:

Fabrics and fibres

Rubber

Cosmetics

Chemicals for farming

Solvents

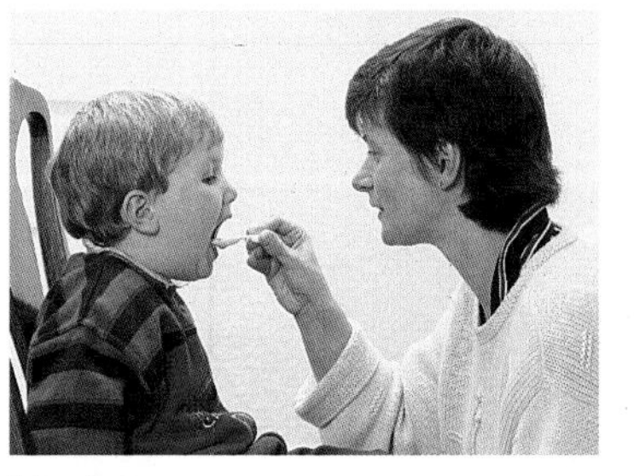

Medicines

Detergents

Plastics

a) Which photo shows a fuel that we get from crude oil?

b) Make a list of useful substances we can make using crude oil as the raw (starting) material.

Crude oil is not a single, pure substance. It is a ***mixture*** of compounds. Most of the compounds are made from hydrogen and carbon. These are called **hydrocarbons**.

Hydrocarbons contain hydrogen and carbon only.

In the beginning ...

The story of crude oil starts about 150 millions years ago. As tiny sea creatures and plants died their bodies collected on the sea bed. These were mixed with other bits of sand and silt that covered the dead animals and plants. They couldn't just rot away because there was no oxygen for bacteria to use down there.

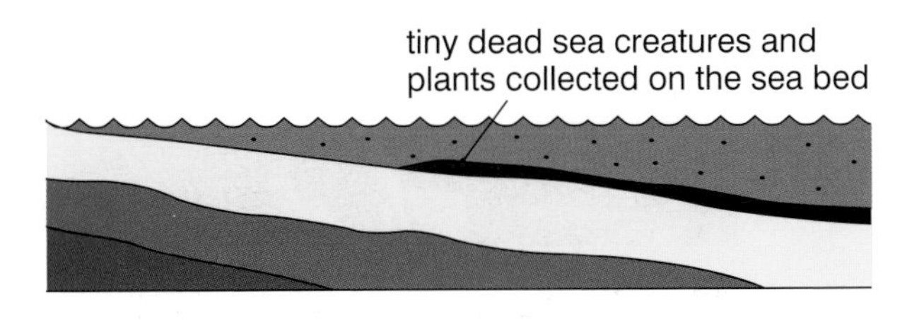

Gradually more and more layers of rock built up. The temperature got higher and so did the pressure, and slowly the crude oil formed.

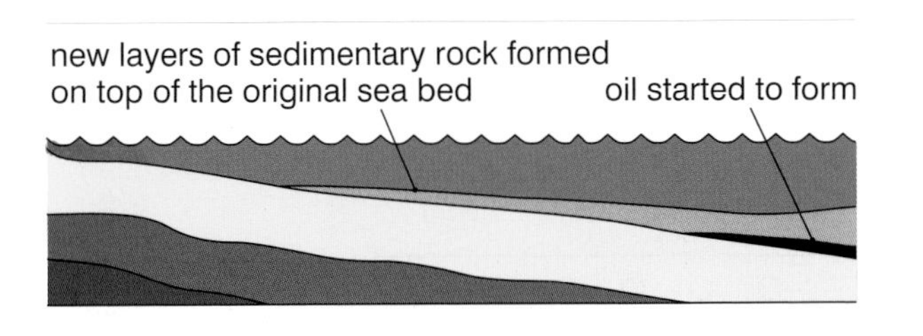

c) What was crude oil made from?

d) Why did the pressure increase as crude oil was forming?

Crude oil is called a **fossil fuel**.
Natural gas and coal are other fossil fuels.
Once we have used up the Earth's supply of these fuels, they can't be replaced.
We call them **non-renewable** fuels.
It is likely that within your lifetime our oil will run out.

Remind yourself!

1 Copy and complete:

Crude oil was formed ...... of years ago from the bodies of tiny ...... creatures and plants.
The crude oil contains a ...... of hydrocarbons.
A hydrocarbon is a compound made up from ...... and ...... only.

2 Do some research to find out:

a) How do we find out where crude oil is?

b) How do we get crude oil to the surface?

c) How is crude oil transported around the world?

d) Which countries produce crude oil?

▸▸▸ 7b Distillation of crude oil

Drilling for oil.

As you have found out, crude oil contains a mixture of hydrocarbons.
When it comes up from below ground or up from beneath the sea bed, it is a thick, gooey and smelly liquid.
The exact mixture does vary depending on where the oil came from.
For example, crude oil from one oil field can be lighter in colour than oil from another place. It probably smells more 'oily' too.

This is because the hydrocarbons are mixed in different amounts.
The various ***hydrocarbon molecules are different sizes***.
Look at some of the smaller molecules below:

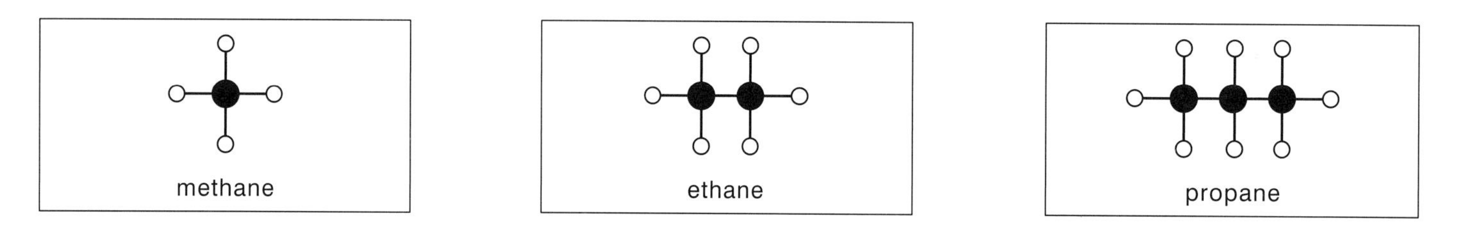

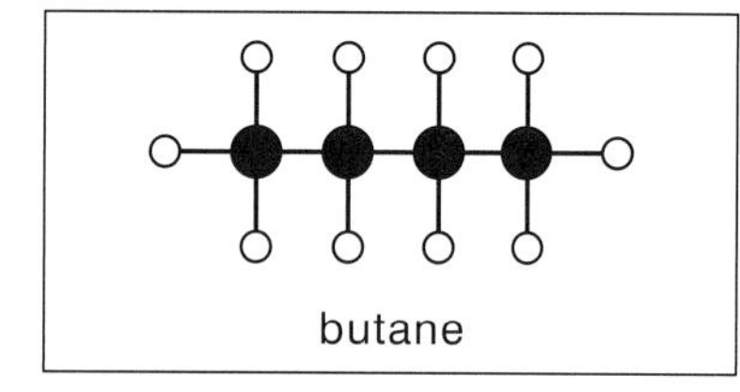

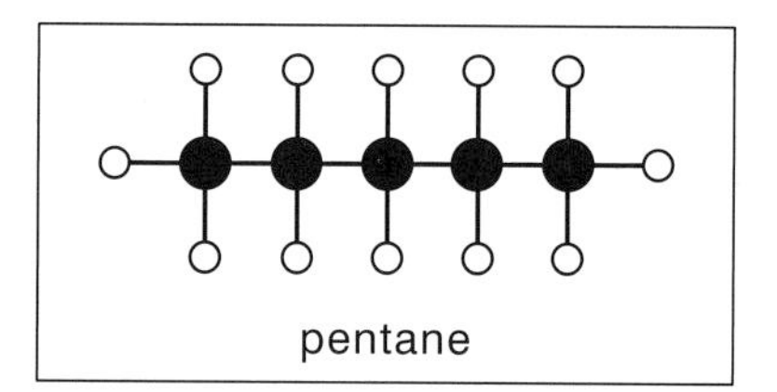

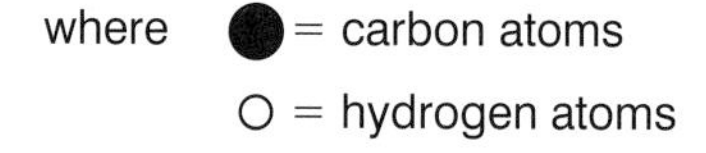

a) Why are the molecules above called hydrocarbons?

b) The chemical formula of methane is CH_4.
What is the formula of butane?

c) What do we use butane for when camping?

All these hydrocarbons, and many other larger ones, are mixed together in crude oil. So oil companies need to separate them to make best use of each type. They do this in their oil refineries. You can read about them on page 82.

To help explain how they do this, look at the experiment on the next page:

Distilling crude oil in the lab

Look at the experiment below:

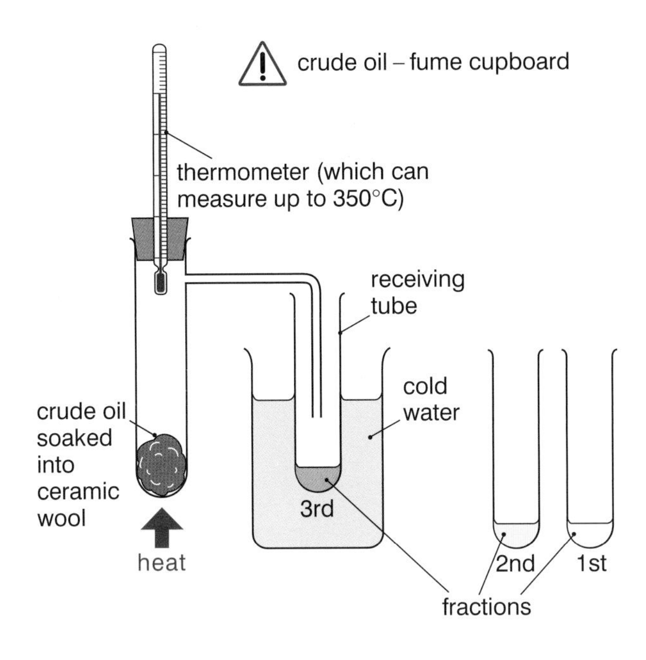

The different hydrocarbons boil at different temperatures.
So we can collect different liquids at different temperatures.
This is called **distillation**.

The liquids are still mixtures of hydrocarbons
– but only those that boil between certain temperatures.
These are called **fractions**.
Look at the table that describes the different fractions below:

Q. What do you call hydrocarbons that tell rude jokes?
A. Crude oil !

Boiling points of fractions	Size of molecules	Colour	Thickness	How it burns
low boiling points (up to 80°C)	small	colourless	runny	lights easily (flammable); burns with a clean flame
medium boiling points (80–150°C)	medium	yellow	thicker	harder to light; some smoke when burning
high boiling points (above 150°C)	large	dark orange	thick (viscous)	difficult to light; smoky flame

Remind yourself!

1 Copy and complete:

The hydrocarbons in …… oil have different boiling points. We can separate them in the lab by …… the mixture.
This gives us groups of molecules with similar …… points called ……

2 Look at the table above to fill in the missing words:

The **larger** the hydrocarbon molecule
- the …… its boiling point,
- the …… volatile it is (the harder it is to evaporate)
- the …… it flows,
- the …… it is to ignite (light).

7c Fractional distillation in industry

Have you ever seen an oil refinery?
These giant sites are a network of huge steel towers, pipes and tanks to store the products from oil.
At the heart of a refinery, the crude oil is separated into its fractions. This is done in huge towers called fractionating columns.
The process is called **fractional distillation**.

An oil refinery at night.

The mixture in crude oil is separated by fractional distillation.

Look at the fractionating column below:

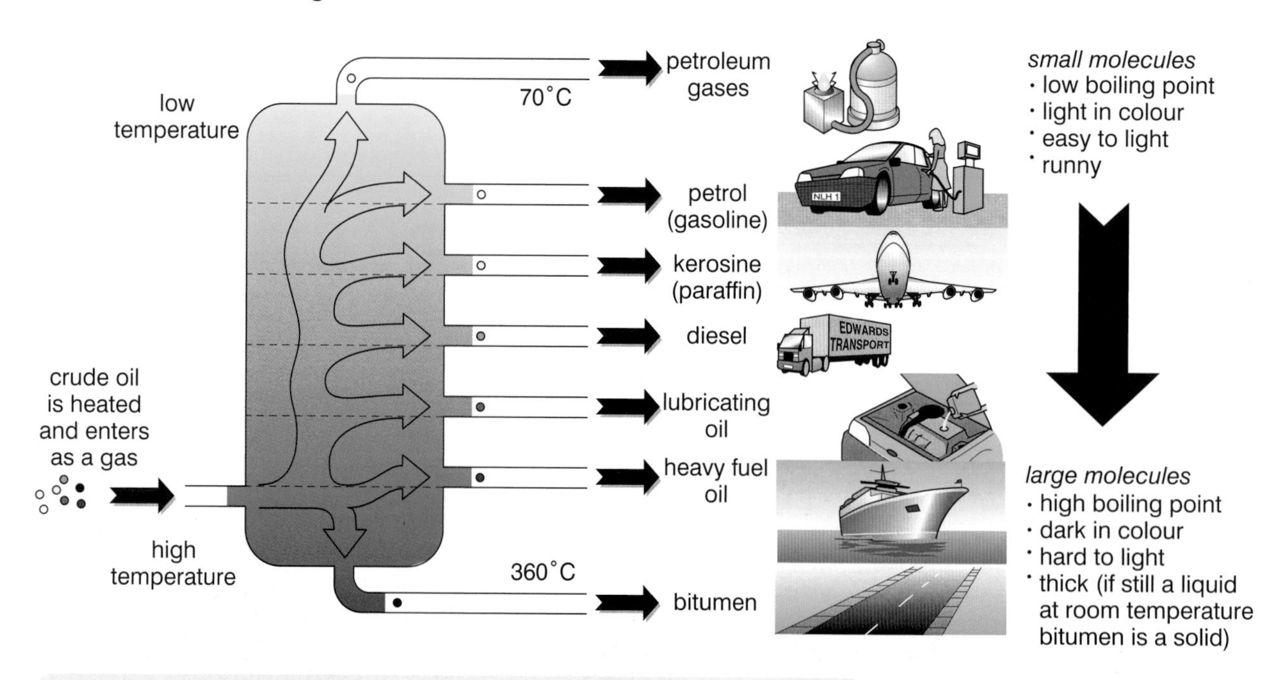

a) Which fraction contains the smallest hydrocarbon molecules?
b) Which fraction contains the largest hydrocarbon molecules?
c) Which fraction shown is not used directly as a fuel?

Explaining distillation

Look at the fractionating column on the last page:

d) Which fraction has the lowest boiling point?

e) Which fraction has the highest boiling point?

f) As the size of a molecule increases, what happens to the boiling point of its compound?

Look at the diagram below:

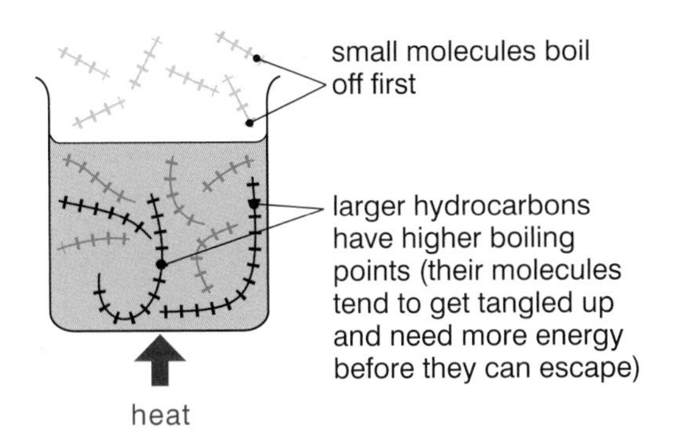

Small hydrocarbons have ***lower*** boiling points than large ones.

The long pieces of spaghetti get tangled up.
They are harder to separate out, just like the long molecules in crude oil.

Remind yourself!

1 Copy and complete:

In an oil refinery, crude oil is separated into its …… by fractional …… The smaller hydrocarbons, with the …… boiling points come out of the …… of the …… column. The largest hydrocarbons, with the …… boiling points, come out at the ……

2 Use the words evaporated or condensed to complete this sentence:

Crude oil is …… before entering the fractionating column and different fractions are …… at different temperatures.

3 Draw a spider diagram showing the products from the fractional distillation of crude oil.

7d Cracking

Imagine you are the manager of an oil refinery:
After your crude oil has been through its fractional distillation, you have a problem.
Your customers want more of the 'lighter' fractions, like petrol. But your crude oil doesn't produce enough.
Instead it contains plenty of the 'heavier' fractions that people don't need so much of.
So what can you do?????
The answer is called **cracking**!

The lighter fractions, such as petrol, are in great demand!

Can you guess what an oil company does with some of its larger molecules in the 'heavy' fractions?
They break them up into the smaller molecules found in the 'lighter' fractions.
And chemists call this 'cracking'.
It certainly is a cracking idea for the oil companies because they make more money, of course!

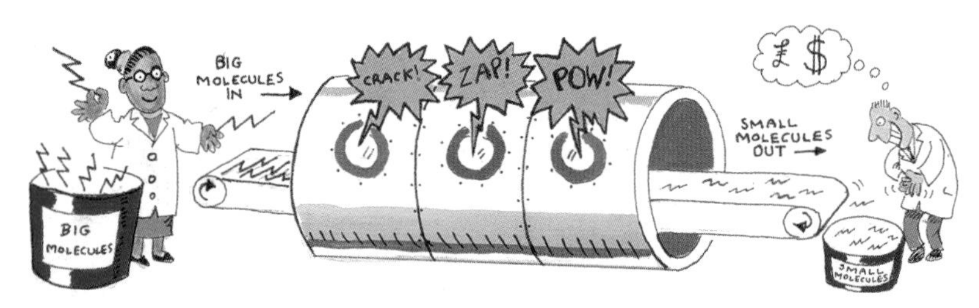

Are these scientists crackers?

Cracking is when we break down large molecules from crude oil into smaller, more useful molecules.

a) When a product is 'in great demand', what does this mean?
b) Why is petrol in great demand?
c) How do oil companies meet the demand for petrol?
Use the word 'crack' in your answer.

small molecules (which include some hydrogen)

large molecules

A cracker in an oil refinery.

In an oil refinery the cracking is done in large steel tanks.
These are called **crackers**.
The large molecules are fed in, then ***heated up***.
A **catalyst** is in the tank. Catalysts make reactions go faster.
(See page 132.)

Cracking in the lab

We can do an experiment ourselves to crack a hydrocarbon.
Look at the diagram below:

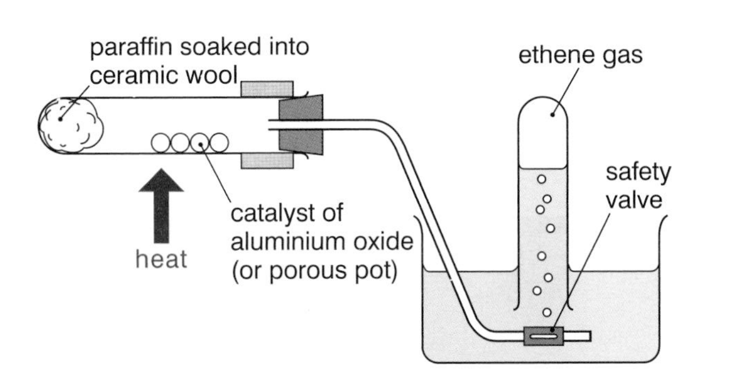

The paraffin evaporates and passes over the hot catalyst.
The paraffin contains large hydrocarbon molecules.
These are broken down into smaller hydrocarbons.

Large hydrocarbons are cracked at high temperatures with a catalyst present.
Cracking is a thermal decomposition reaction.

d) What is the catalyst in the experiment above?
e) What does a catalyst do in a reaction?
f) Name the gas formed in the experiment.

It's a cracker!

Remind yourself!

1 Copy and complete:

In an oil refinery, we get more of the ...... fractions than we need. We need more of the ...... fractions, such as ......
To meet the demand, oil companies ...... larger hydrocarbon molecules into smaller, more ...... ones. They do this in steel vessels called ......
The large molecules are ...... up and passed over a ...... to break them down.
This is a ...... decomposition reaction.

2 One of the compounds in petrol is called octane. Its chemical formula is C_8H_{18}. You can get it by cracking a larger hydrocarbon called decane. Ethene, whose formula is C_2H_4, is formed as well as octane.

a) Write a word equation to show the cracking of decane.

b) Write a balanced equation for the cracking of decane.

▶▶▶ 7e Burning fuels

Our lives today rely on the fuels we get from crude oil. Remember that we separate these from the oil by fractional distillation. We make more by cracking larger, less useful hydrocarbons from the distillation.

Burning fuels

How would you put out a fire in a chip pan?
The damp tea towel stops oxygen getting to the burning oil.
But if you take the tea towel away too soon,
the hot cooking oil can burst back into flames.
If cooking oil catches fire outside on a barbecue,
you can just let the fire burn harmlessly.
It stops when all the cooking oil is used up.

A chip pan fire.

a) Think about the burning cooking oil:
Which 3 things are needed for it to burn?

We need 3 things for a fire.
This can be shown in the fire triangle:

Fuel — Oxygen — Heat

Whenever you come across **heat**,
And **fuel** and **oxygen** meet,
You're sure to get fire,
The flames will lick higher,
The **fire triangle**'s complete!

Look at the photo opposite:
It shows a fire at a gas rig in the North Sea.

b) How would you put out a fire at a gas or oil rig?
Which part of the fire triangle has been removed?

c) What was needed to start the fire on the gas rig?

d) How would you put out a fire in a waste paper bin?
Which part of the fire triangle are you removing?

What is formed when fuels burn?

Young children often think that when a liquid fuel burns it just disappears. Of course, we know that's not true! The chemical reaction, called **combustion**, gives off gases.

We can test which gases are made in the experiment below:

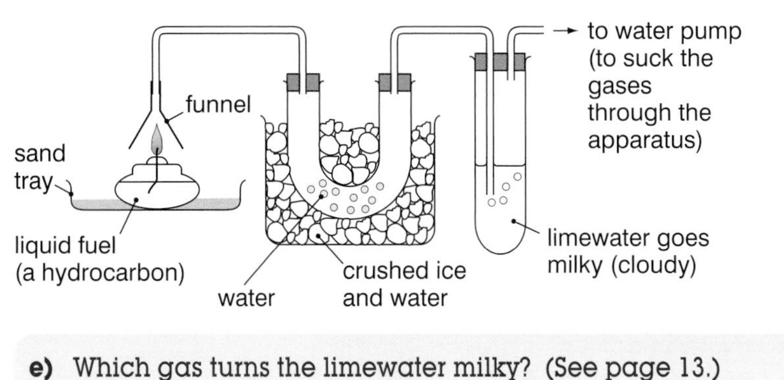

e) Which gas turns the limewater milky? (See page 13.)

f) What is formed in the U tube?

The equation for burning a hydrocarbon is:

hydrocarbon + oxygen → carbon dioxide + water

When the fuels we get from crude oil burn, we get:

- **Carbon dioxide gas**
- **Water (vapour)**
- **Sulphur dioxide gas**

Inside an engine, there is not much oxygen. So some ***carbon monoxide*** *gas is formed. This is a toxic gas with no smell. It stops your blood carrying oxygen around your body.*

So where does the sulphur to make sulphur dioxide come from? The sulphur is an impurity in the fuels we get from crude oil. You can find out about the pollution caused by burning hydrocarbons on the next page.

Remind yourself!

1 Copy and complete:

Three things are needed for a fire – ……, …… and ……
The fuels we get from crude oil produce carbon ……, …… vapour and …… dioxide when they burn. The …… comes from impurities in the crude oil.
If there is insufficient …… we also get toxic carbon …… formed.

2 Octane is a hydrocarbon in petrol.

a) What is formed when pure octane burns in plenty of oxygen?

b) Untreated petrol has some sulphur in its mixture. What gas does the sulphur form when it burns?

c) Find out about lead in petrol and why it has been banned.

7f Pollution from fuels

Are you worried about **air pollution**?
You probably notice the problem more if you live in a town, especially if you also have asthma.
Most of the air pollution comes from burning fuels.

In Tokyo you can get oxygen in special bars. This fights the effects of pollution from cars in the busy streets.

Whenever you travel very far you burn fuels from crude oil to get there. Imagine this holiday:
You get a taxi to the train station, then catch a train to the airport. You fly off to the Caribbean to join a cruise ship for two weeks.

a) Which fuel do you use on each stage of your trip? (Look back to the fractions on page 82.)

Even when you turn on a light switch, you are probably using up a little more fuel. That is true if your electricity is generated in a power station that burns coal, oil or gas.

b) Name a source of electricity that does not rely on burning fuels.

Coal-fired power stations give off sulphur dioxide gas. This causes acid rain.

Acid rain

As we saw on the previous page, **sulphur dioxide gas** is given off when we burn some fuels.
This causes acid rain. The sulphur dioxide dissolves in the tiny water droplets in clouds.
This forms a solution of sulphuric acid.

When the acid rain falls it causes the problems below:

- Forests – trees are damaged and killed.
- Fish – aluminium, which is normally held in the soil, is washed into lakes and rivers. The aluminium poisons the fish.
- Buildings – acid rain attacks buildings, especially made from limestone, and metal structures.

To help reduce this problem we should:
– use 'low' sulphur fuel in our cars,
– save electricity whenever possible.

Governments are also funding more research into other sources of electrical energy that don't burn fossil fuels.
The acidic gases from power stations can also be 'cleaned'.

The greenhouse effect

For a long time people thought that another gas given off from burning fuels was perfectly harmless. **Carbon dioxide** is in the air naturally (about 0.04%). But now there is concern about **global warming.**

The average temperature of the Earth is slowly rising. Most scientists agree that the increasing amount of carbon dioxide in the air is the main cause. Some people blame this for unexpected disasters caused by freak weather (like flooding from heavy rain).

Trees absorb carbon dioxide during photosynthesis. But huge areas of rain-forest are being cut down every year.

In the last 100 years we have been greedily burning more fossil fuels than in the rest of history put together. Nature has its own ways of removing carbon dioxide from the air, but there is just too much being made now. We have disturbed the natural balance.

c) How do plants remove carbon dioxide from air?

d) Why does cutting down rain-forests make global warming happen faster?

We call carbon dioxide a 'greenhouse' gas. Look at the diagram opposite:

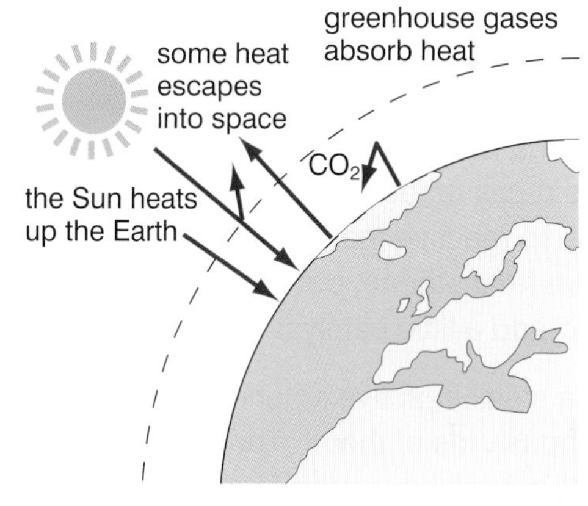

Carbon dioxide and water vapour are the main 'greenhouse gases'.

The Sun heats the Earth. It cools down by giving off infra-red (heat) rays. But carbon dioxide absorbs the heat, so it can't escape out to space. Heat gets trapped – like in a greenhouse. So people call it the **greenhouse effect**.

To help reduce the problem, as with acid rain, we must burn less fossil fuels. We should also plant as many trees as possible to absorb carbon dioxide.

New cars have catalytic converters fitted in their exhausts. These help with reducing acid rain as they get rid of acidic nitrogen oxides. They also get rid of toxic carbon monoxide, but they change this to carbon dioxide. So the catalytic converters don't help reduce global warming!

Rising sea levels
People are worried about the level of the oceans rising because of global warming. Polar ice caps could melt and the hotter seawater will expand.

Remind yourself!

1 Copy and complete:

Burning fuels from crude oil causes pollution of the ...... For example, ... dioxide causes acid ...... This kills trees and ......, as well as damaging metals and ......
...... dioxide is the main cause of the ...... effect, which produces ...... warming.

2 In 2001 Britain signed an international agreement to reduce the amount of carbon dioxide released from each country. But some other countries refused to sign.

Write a letter to a newspaper explaining why it's so important to reduce the CO_2 we produce.

7g Plastics

We take plastics for granted as part of modern living.
Just think of all the plastic things you use each day.

How many things in this room are made of plastics or synthetic fibres?

a) List 4 things made of a plastic that you have used today.

You know that plastics are synthetic.
But do you know where we get the raw materials to make them from? It might not be obvious, but most come from products we get from crude oil.

Do you remember the work we did on cracking (page 84)?
One of the products we get when we crack a large hydrocarbon is called **ethene**.

b) What is a hydrocarbon?

Ethene is a small hydrocarbon. Its formula is C_2H_4.
It is a gas made of very reactive molecules.
It is so reactive that it can even react with itself!
If we heat ethene, put the gas under pressure, and add a little catalyst – Hey presto! We get a plastic.

The small, reactive ethene molecules join together – thousands of them! They make really big molecules.
In this case we get poly(ethene) formed.
You probably know it better as polythene.

Low-density poly(ethene) was used to make this bag.

Lots of ethene molecules → poly(ethene)

c) What does 'poly' mean when we put it in front of a word?

The small, reactive molecules that join together to form the big molecule are called **monomers**.

The big molecule formed is called the **polymer**.

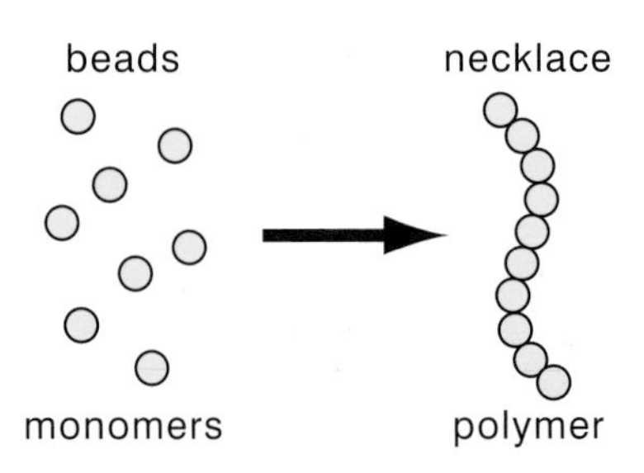

d) What is the monomer in the reaction described on this page?
e) What is the polymer in the reaction above?

The reaction of monomers to form a polymer is called **polymerisation**.

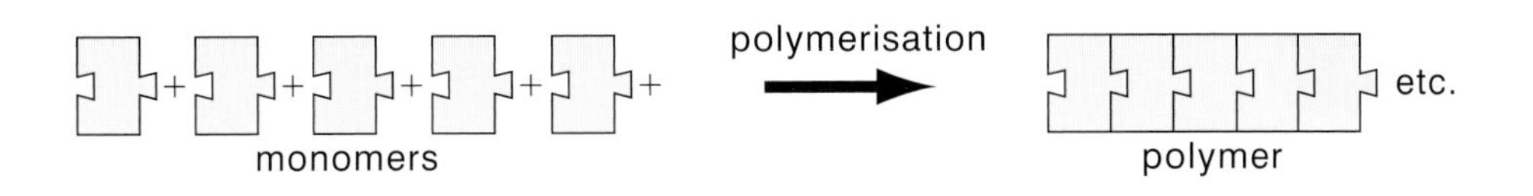

Lots of monomers → polymer

Another monomer we use to make plastics is propene.

Lots of propene molecules → poly(propene)

f) Name the polymer made from the monomer called styrene.

Look at the uses of poly(ethene) and poly(propene) below:

Remind yourself!

1 Copy and complete:

Small, ... molecules that we get by ...... large hydrocarbons can be used to make ......
For example, ... makes poly(ethene), and propene makes ......

The small molecules are called ...... and the large molecule made is called a ......

2 Plastics have only been in common use for less than a hundred years.

What were these objects made from before plastics and polymers took over?

a) gutters

b) wrapping for food

c) kitchen work-tops

d) pens.

7h Getting rid of plastics

Just take a look at the amount of plastic you throw away after shopping in a supermarket!
We use plastics because they are cheap. We can mould them into any shape or draw them out into fibres and very thin sheets.
These are all very useful properties.

We generate a lot of plastic waste.

a) How do you think that plastics are moulded?

Plastics are also very unreactive.
Of course, we don't want plastic gutters that will leak as soon as water flows down them. Imagine a plastic duck that dissolved away when you put it in the bath!
But it is a plastic's lack of reactivity that causes us a problem.

What happens when you are fed up with your plastic duck?
You throw it away and it will probably end up at a land-fill tip.
However, many plastics take a ***very*** long time to rot away.
The micro-organisms in the soil cannot break it down.
We are running out of space at these rubbish tips, so plastics take up valuable room.

A land-fill rubbish tip.

b) What problems would a council have when wanting to build a new land-fill rubbish tip?

So scientists are working on ways to help plastics rot away once we dump them.
These are called **biodegradable plastics**.

Plastics that rot when we throw them away are called biodegradable.

c) What is the main advantage of a biodegradable plastic?

Bottles made from biodegradable plastic shown as they break down.

Re-cycling plastics

Do you re-cycle paper, glass or aluminium cans?
More and more people are trying to help save our resources.
But re-cycling plastics is more difficult.
There are so many different types.
But other countries do much better than Britain.
Look out for these symbols on plastic goods:

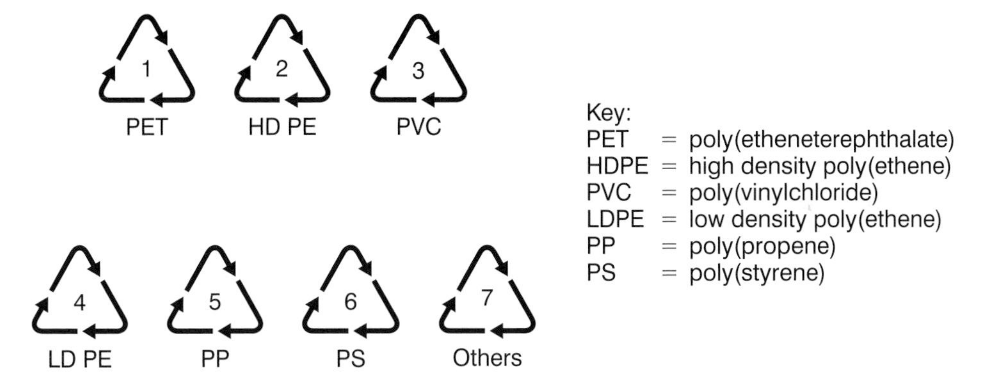

These can be melted and re-moulded for new uses.
This saves our supplies of crude oil (the raw material for plastics) and saves space in land-fill sites.

d) Why is recycling plastics more difficult than glass?

Burning plastics

Many plastics are hydrocarbons. As you know, hydrocarbons are our main source of fuels.
So perhaps we could burn waste plastics and use the energy released instead of using up fossil fuels?
Look at the bar chart opposite:

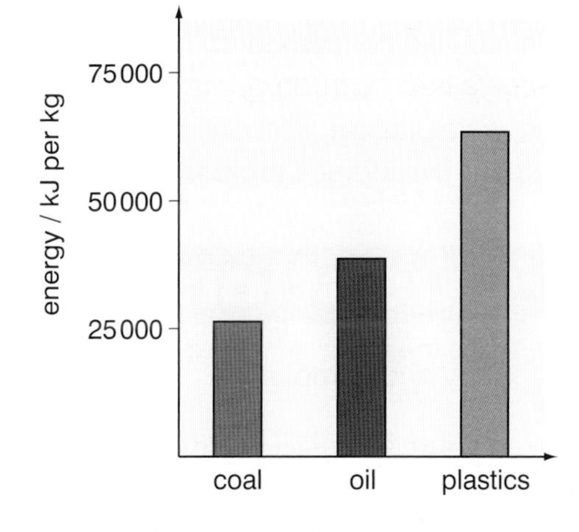

But there are problems with this idea.
Some plastics give off toxic gases when they burn.

e) How could we solve the problems with burning plastics?

Remind yourself!

1 Copy and complete:

Plastics have many useful properties but their lack of ...... is a problem.
They do not ...... when we dispose of them and are taking up valuable space in ...... sites.
However, more plastics made now are ......

2 List as many ways as you can that you can help solve the problem of disposing of plastic waste.

3 Do some research to find out more about plastic waste and how we are tackling the problem.
Present your findings in a report to send to your local council.

Summary

Crude oil is a mixture of hydrocarbons.
It can be separated into compounds with similar boiling points by **fractional distillation**.

Most of the fractions from crude oil are used as fuels.
Fuels such as petrol are in great demand.
So some large hydrocarbon molecules, the heavier fractions, are **'cracked'** into smaller, more useful molecules to use as fuels.

My name is methane –
If you want a fire,
Just light the gas
And turn me higher!

During cracking, we also get small reactive molecules such as ethene formed.
These can react with each other when heated under pressure in the presence of a catalyst. They join together to make very large molecules used to make plastics.
The small molecules are called **monomers**.
The large molecule they form is called a **polymer**.

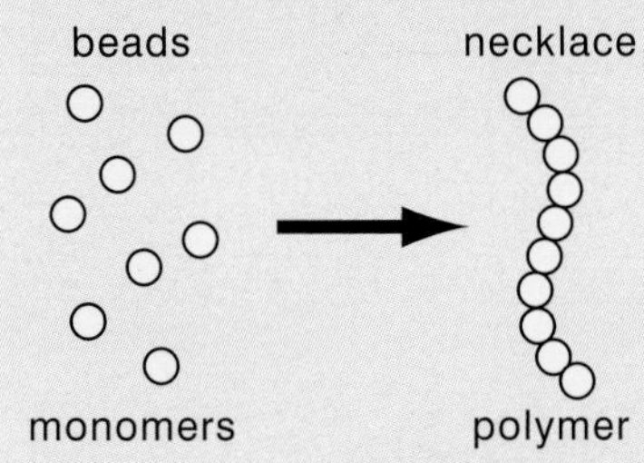

There is a problem disposing of plastics because when we throw them away many are not broken down by micro-organisms in the soil.

We are also concerned about air pollution caused by burning fossil fuels.
When a hydrocarbon burns we get carbon dioxide (**greenhouse effect**) and water vapour. Impurities of sulphur in the fuels also produce sulphur dioxide gas (**acid rain**).

Questions

1 Copy and complete:

There is a …… of different hydrocarbons in …… oil.
We can separate out more useful mixtures with …… boiling points by …… distillation.
We can …… larger hydrocarbon molecules into ……, more useful molecules. In this process, small, …… molecules such as …… are also formed. These can be used to make ……
Thousands of …… join together to form a large molecule called a ……
…… in the soil cannot break down many plastics, so we are developing more …… plastics.

2 Sketch this diagram of a fractionating column:

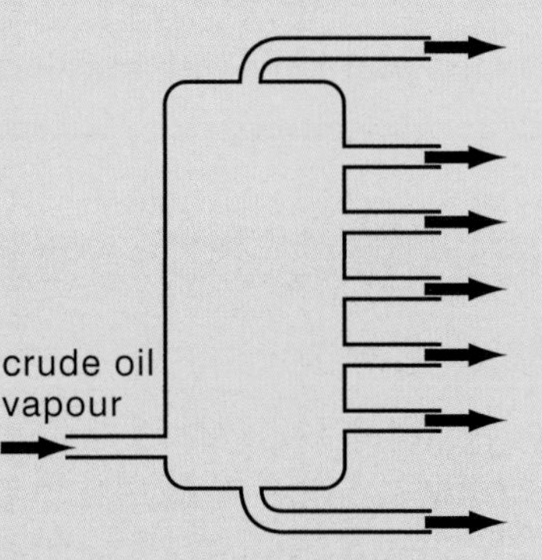

Fill in the missing fractions.
Say what each fraction is used for.

3 In an experiment to distil crude oil in the lab, three fractions were collected. But they were not kept in the order that they boiled off.

a) Give 4 ways that you could tell which fraction contained compounds with the lowest boiling points.

b) In an oil refinery, from which part of a fractionating column do we draw off the fraction with the lowest boiling points?

c) What is a hydrocarbon?

d) What can you say about the size of a hydrocarbon molecule and its boiling point?

4 Look at the experiment below:

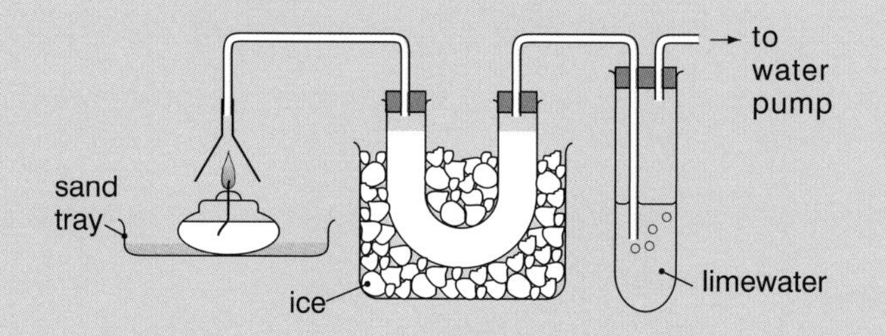

a) Why do we put ice around the U tube?

b) One test for water is that it turns blue cobalt chloride pink.
Give another chemical test for water using anhydrous copper sulphate.

c) What is the missing number below:
You could check that the liquid collected in the U tube was pure water by testing if its boiling point is ......°C.

d) What happens to the limewater in the experiment?

e) Which gas do we use limewater to test for?

f) Write a word equation for burning a hydrocarbon in plenty of oxygen:

hydrocarbon + oxygen → ...... + ...

g) If a fuel contains sulphur impurities, which gas is made when it burns?
Which pollution problem does this gas cause?

5 a) Complete the table below:

Monomer	Polymer
ethene	......
......	poly(propene)
styrene	......
......	poly(vinylchloride)

b) Draw a table showing 3 uses of poly(ethene) and 3 uses of poly(propene).

c) Describe some of the ways that we can help to solve the problem of disposing of plastic waste.

6

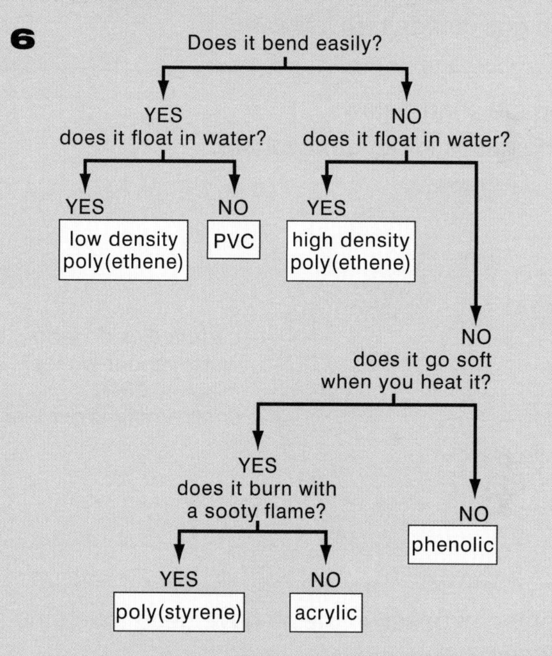

a) Which plastic bends easily and sinks in water?

b) Which plastic does ***not*** bend easily, nor float in water, nor go soft when you heat it?

c) Use the key to say as much as you can about acrylic.

7 Imagine you are a small hydrocarbon molecule in crude oil.

Describe what happens to you from the time you are discovered to the time you end up heating beans on a camping stove.

8 Design a leaflet to inform Y7 pupils about the greenhouse effect. It should tell them what it is, the effect on the environment, and how we can help stop it.

CHAPTER 8 AIR and ROCKS

8a Gases in the air

Every time you breathe in, you inhale a mixture of gases that make up air.

a) Which gas couldn't we live without?

Look at the pie chart below:
It shows the gases in our atmosphere.

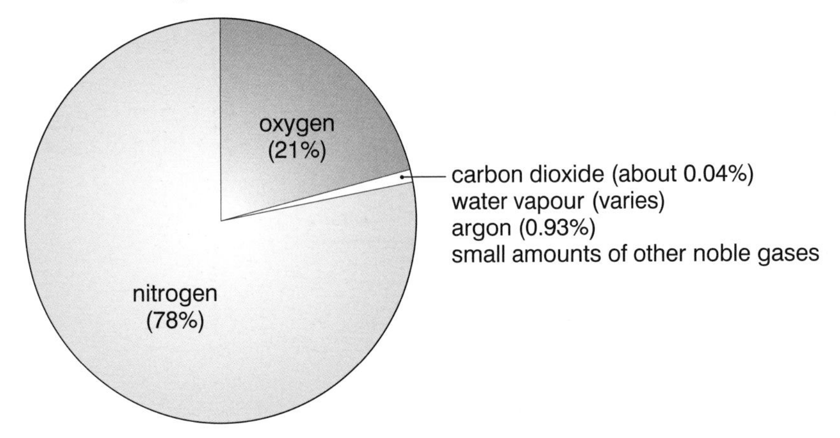

The proportion of gases in the air have been much the same for the last 200 million years!

b) Which gas makes up most of the air?
c) Which gas is taken in by plants during photosynthesis?
d) What is the missing word?
Argon is a n...... gas.

Nitrogen

Nitrogen is a very un-reactive gas.
That's why we use it inside food packaging.
Without oxygen the bacteria can't decompose the food.

It is also used when we unload the oil off giant tankers.
If there is a spark, the ship could explode.
So nitrogen gas is pumped in to get rid of any oxygen.

Nitrogen is used to avoid the risk of explosions when pumping oil ashore.

e) What mixes with oxygen to form an explosive mixture in an oil tanker?

Water in the air

You can see from the pie chart on the last page that the amount of water vapour in the air varies.

f) Name two places where you would expect very different amounts of water vapour in the air.

g) What do we mean when we say that it's a humid day?

Look at the **water cycle** below:

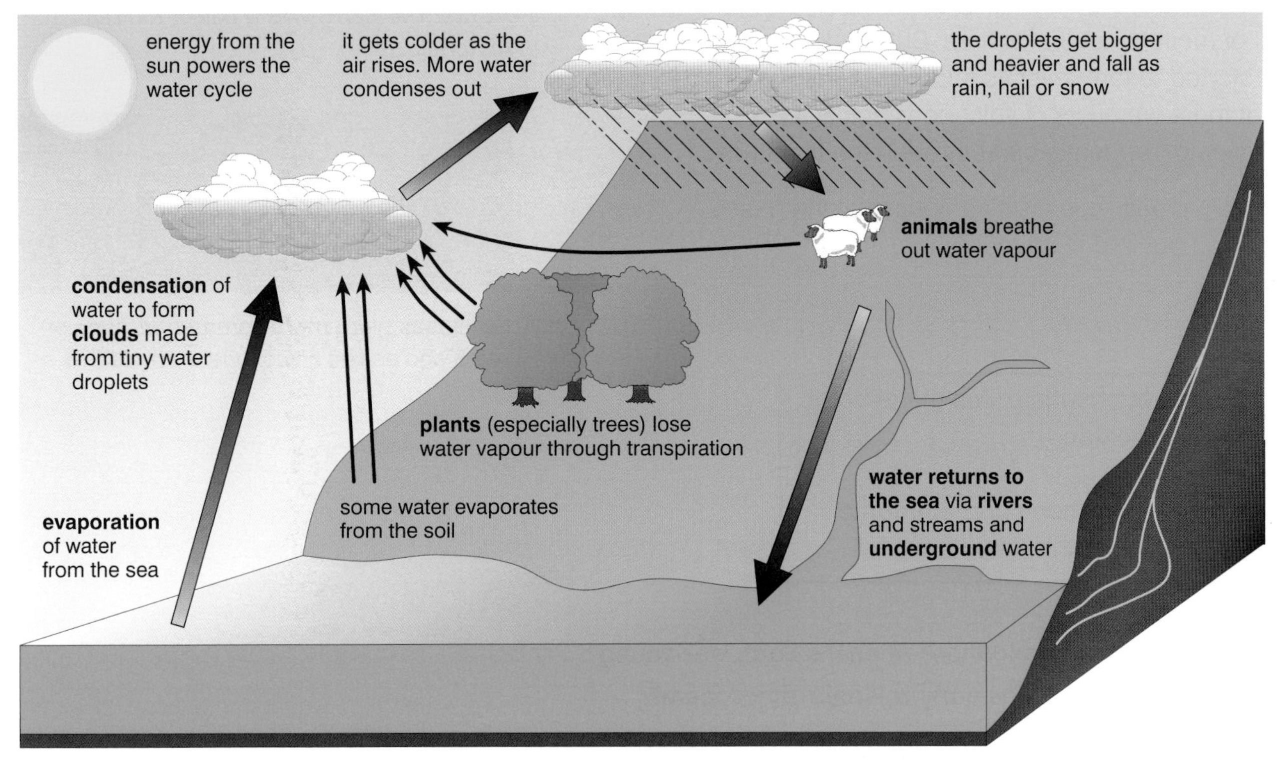

The water cycle shows us how the world's water moves from place to place.

h) Where does the energy come from to drive the water cycle?

Remind yourself!

1 Copy and complete:

For the last …… million years, the Earth's …… has remained almost the same.
The main gas in the air is ……, with …… the second most common gas. The proportion of …… dioxide in the air is about 0.04%.
The amount of water vapour …… at different times and places.

2 Look at the water cycle above:

a) What do we call the process whereby trees lose water vapour?

b) Why do the water droplets get bigger as clouds rise?

c) Draw a flow diagram to summarise the water cycle.

3 Do some research to find out how oxygen gas was discovered.

▶▶▶ 8b History of the atmosphere

Have you ever wondered where the gases in the air came from? Scientists have spent a lot of time thinking about this. They have put forward their ideas, although there is still some disagreement. After all we are talking about things that happened many millions of years ago.

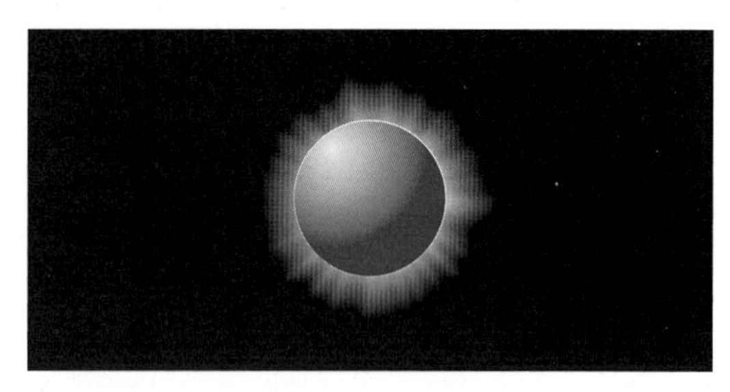

At first the Earth was a ball of molten material.

Most scientists believe that the Earth was formed about 4.6 billion years ago. (That's 4 600 000 000 years ago!)

It was a ball of molten material at first. But gradually it cooled down.
A thin crust started to form over its surface.
We think that there were lots of volcanoes at this time.
Molten rock would frequently burst through the thin crust.

Volcanoes were more common millions of years ago as the atmosphere started to form.

By studying volcanoes, scientists can guess at the gases that also spewed out into the atmosphere.
They think that:

> The early volcanic atmosphere was made up of:
> - mainly **carbon dioxide**,
> - **water vapour**,
> - small amounts of **methane** and **ammonia**.
>
> There was ***no oxygen*** at this time.

a) Why were there so many volcanoes when the Earth was young?
b) What was the main gas in the early volcanic atmosphere?

As the Earth carried on cooling down, the water vapour condensed and fell as rain. The rain filled the hollows in the Earth's crust and the first oceans were formed.

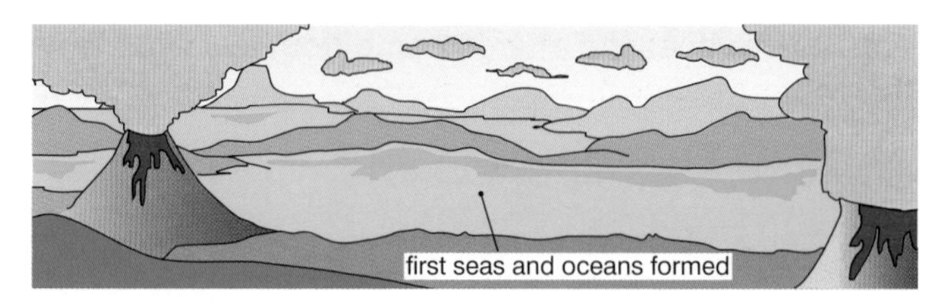

As the Earth cooled, rain fell and formed the oceans.

This was important because the first living things developed in the sea.
Recently scientists have found all the elements needed for life around volcanic outlets at the bottom of the ocean.
The simple organisms evolved into small ***plants***, like algae.

The first oxygen now appeared on Earth.
During **photosynthesis**, carbon dioxide was taken in and oxygen was given out. As plants spread over the Earth, other living things that could not tolerate oxygen died out.

c) How did the first oxygen get into the atmosphere?

The levels of carbon dioxide slowly decreased as the oxygen increased.
Carbon from the carbon dioxide gradually became 'trapped' in ***fossil fuels***. For example, trees and ferns that died in pre-historic swamps formed coal.

Carbon from the pre-historic atmosphere is 'trapped' in coal.

Carbon dioxide was also removed by dissolving in the oceans. This carbon eventually turned up in ***carbonate rocks***, some formed from the shells of sea creatures. For example, some types of limestone are formed from shells that were covered by layers of sand and mud. It is called a sedimentary rock. (See next page.)

d) Name 3 fossil fuels.
e) Name a sedimentary carbonate rock.

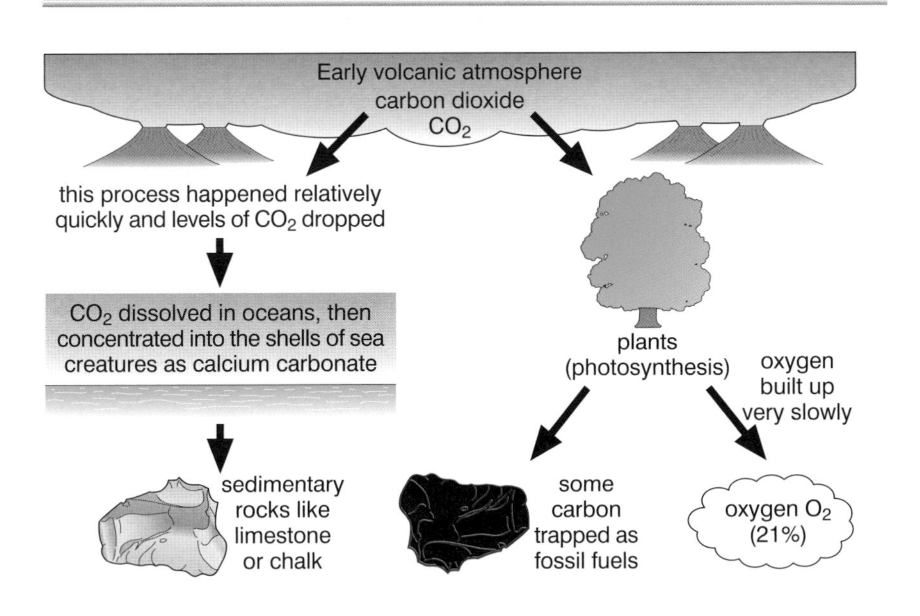

The methane and ammonia were removed from the early atmosphere by reacting with oxygen.
So eventually the air reached the balance of gases that we have in today's atmosphere.

Remind yourself!

1 Copy and complete:

The early volcanic atmosphere was made up mainly of …… …… gas.
There was also …… vapour (which cooled down to form the ……) and small amounts of …… and ammonia.
The first …… was produced by plants.
Much of the carbon from carbon dioxide became trapped in …… fuels and carbonate ……

2 a) Find out how coal was formed.

b) How did carbon dioxide get into the early atmosphere?

c) Explain how some of this carbon from CO_2 is now trapped in fossil fuels.

d) How does the carbon trapped in fossil fuels get back into the atmosphere?

e) Why are people worried about the levels of carbon dioxide building up in the atmosphere?

8c Sedimentary rocks

Do you know what the word 'sediment' means?
If you know about home brewing, you'll recognise the sediment that collects at the bottom of the flask.
Sedimentary rock forms from bits of rock and other debris that build up over time.

There are 3 types of rock:
- **sedimentary,**
- **metamorphic,**
- **igneous.**

Sandstone is a sedimentary rock.

So where does the sediment come from to form the sedimentary rock? On the previous page we saw how some types of **limestone** were formed. The shells of sea creatures collected on the sea bed. These were covered by layers of more shells, sand and mud. Under this pressure (compaction), the water in between the individual particles was squeezed out. This leaves behind any minerals that were in solution. These help to 'cement' the particles together.

a) What is the sediment that forms the main part of limestone?
b) Which carbonate makes up most of limestone? (See page 68.)

Other sedimentary rocks are formed from different bits of eroded rock. These were usually carried along by rivers, then deposited on ancient sea beds. Over millions of years the layers build up and the processes of compaction and cementing take place.

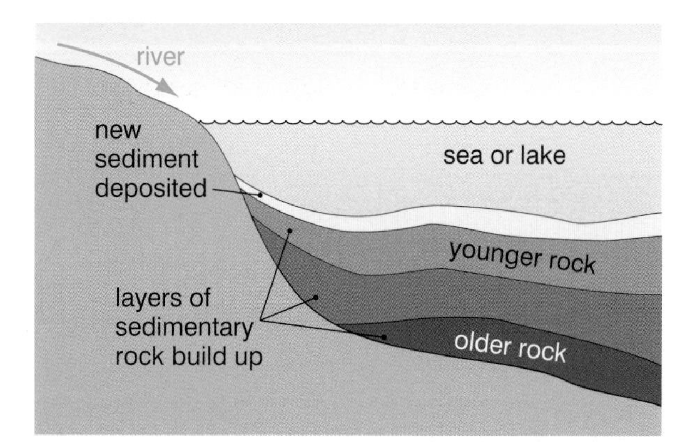

Layers of sedimentary rock forming.

c) What does 'compaction' mean?
d) In several layers of sedimentary rock, where do you normally find the oldest rock?

Look at the different types of sedimentary rock below:

Particles	Sedimentary rock
clay ⟶	mudstone
sand ⟶	sandstone
pebbles ⟶	conglomerate

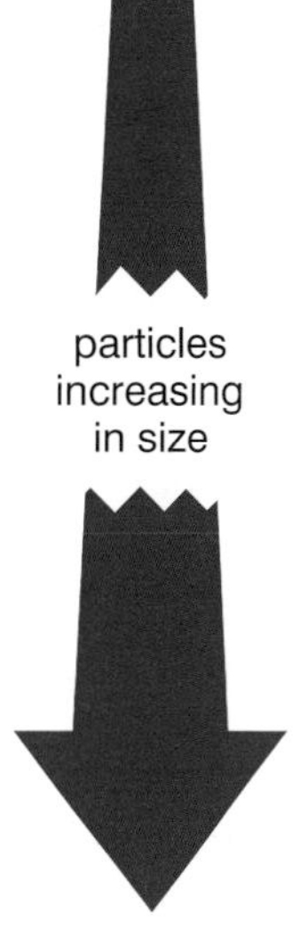

Other sedimentary rock was formed when ancient seas evaporated, leaving behind the dissolved salts.
For example, the seam of rock salt beneath Cheshire is up to 2000 metres thick in places.

e) Which sedimentary rock is formed from clay?

Coal is another sedimentary rock.
Like limestone, you often find fossils in coal.

f) How was coal formed?
g) What type of fossils would you expect to find in limestone?
h) What type of fossils would you find in coal?

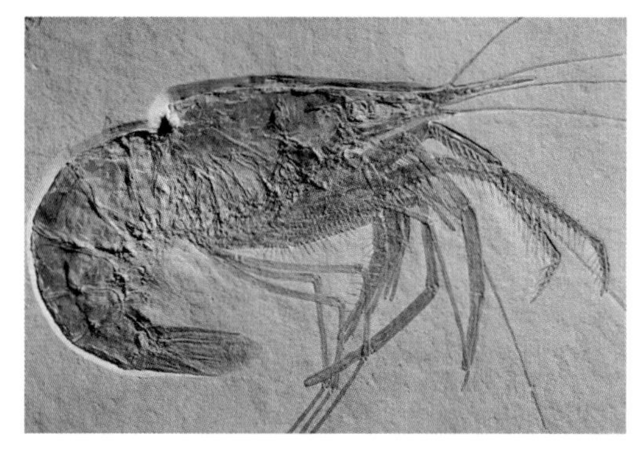
Fossils are often found in sedimentary rocks.

Fossils can help us to 'age' or 'date' the rocks they are found in.
More advanced animals and plants are found in younger rock. We find simpler, more primitive species in older rock.

Remind yourself!

1 Copy and complete:

Rock is weathered and then ...... along by wind or in ...... The bits are ...... on the ...... bed where they slowly get ...... by more and more layers.
The particles get c...... under pressure as the layers build up. Water is squeezed out and minerals c...... the particles together. They then form ...... rock.
We often find ...... in this type of rock.

2 These two fossils were found in two different places.
Which fossil was found in the younger rock?

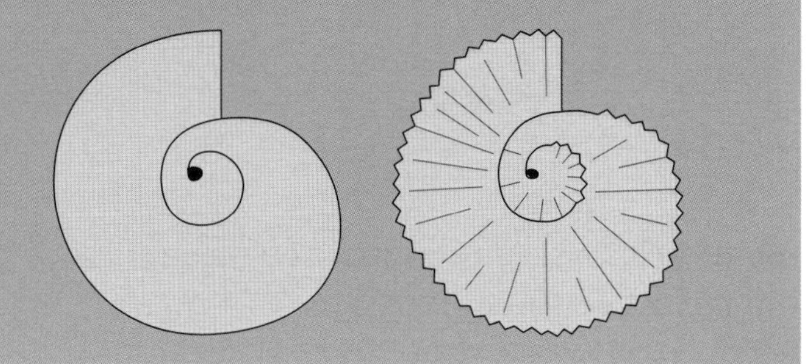

8d Metamorphic rocks

Do you know of any houses with **slate** roof tiles in your area? About a hundred years ago, slate was the most popular material to use on roofs. It is a rock that can be split into thin sheets to make the tiles.
Slate is an example of a **metamorphic** rock.

Slate is a metamorphic rock. It is formed from mudstone.

You might have heard of the word 'metamorphosis' in your biology lessons. An example would be when a caterpillar changes to a butterfly.
The animal has changed its form.
So a metamorphic rock is one that changes its form.
But how can that happen?

In the next chapter we will see how the Earth's crust can move – slowly, but it can move.
This means that rocks can be buried deep underground, and are squeezed when two parts of the crust collide and form mountains. The rocks can be put under ***great pressure***.
Look at the diagram opposite:

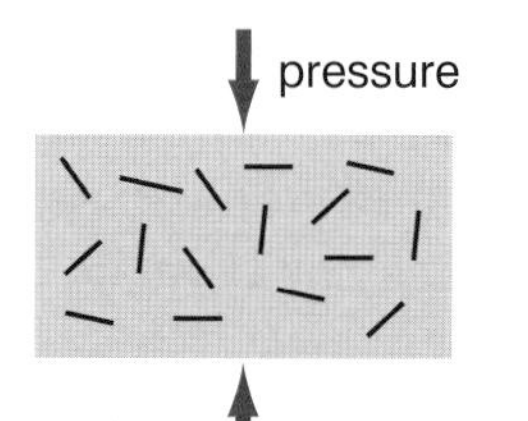

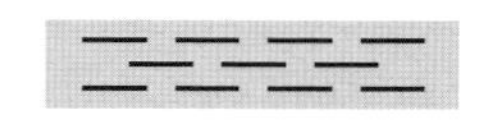

Rocks can also be heated to very ***high temperatures***.
Inside the Earth there is molten rock called **magma**.
This rises towards the surface through any cracks in the crust.
The rocks next to the red hot magma get baked.
They don't actually melt, but their structures are changed.

Look at the diagram below:

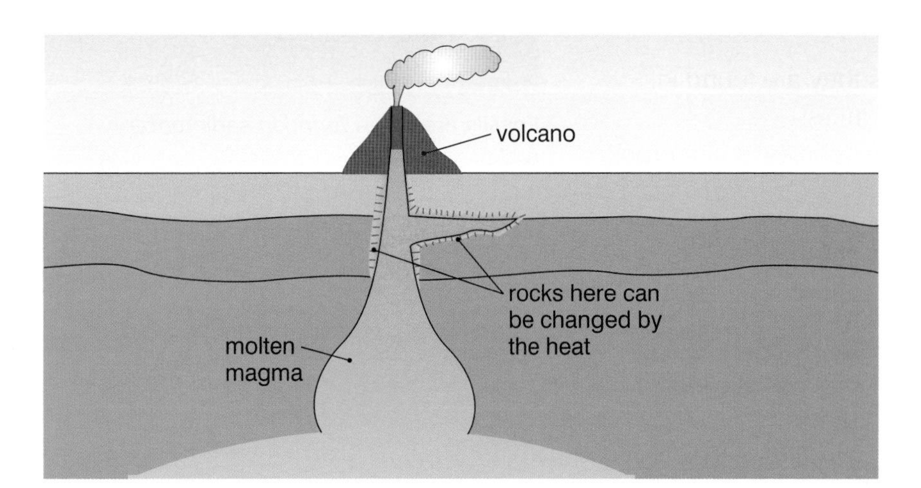

a) How can the structure of a rock be changed?

b) Name two metamorphic rocks.

In some metamorphic rocks, you can see bands running through the rock.
The bands will be at right angles to the direction of the pressure.
Look at the rock, called **gneiss**, below:

Gneiss is a metamorphic rock.

c) Sketch the gneiss above and include arrows to show how the rock was squeezed.

Marble is another metamorphic rock.
It is made when limestone is baked at high temperatures deep underground.

Marble is formed from limestone.

Metamorphic rocks: These are rocks that have been changed by pressure and/or heat (without melting).

Remind yourself!

1 Copy and complete:

The Earth's crust continues to ...... slowly and rocks can be subjected to very high ...... as they are squeezed. This can force the minerals to l...... up and form bands. Rocks can also have their structures ...... by high temperatures. The new rocks formed are called ...... rocks.

2 a) Which sedimentary rock formed slate?

b) Explain how slate was formed.

c) Why is slate used on roofs?

3 Explain how a seam of limestone beneath the Earth's surface can be heated to a high temperature. Which metamorphic rock is formed?

8e Igneous rocks

Do you know of any word that begins with ign......
(other than igneous!)?
The most common word is ignite – to set something on fire.
This gives us a clue how igneous rocks were formed.

Igneous rocks are hard!

> **Igneous rocks** are formed when molten rock cools down.
> They are made up of ***crystals***.

Look at the examples below:

Granite is an igneous rock.

Gabbro is another igneous rock.

a) Are the rocks above mixtures or pure substances?
b) Do you think that igneous rocks are hard or soft?

The molten rock comes from deep inside the Earth.
When molten, it is called **magma**.
As it rises towards the surface, it cools down.
The crystals form as the rock turns into a solid.

c) What happens to the temperature underground as you get deeper and deeper?

The size of the crystals in an igneous rock depends on how quickly the molten rock cooled down when it formed.

If an igneous rock cooled down ***quickly***, its crystals are ***small***.
If an igneous rock cooled down ***slowly***, its crystals are ***large***.

Basalt has small crystals.

d) Name an igneous rock that cooled down quickly as it formed.

So why does some molten rock cool down quickly and other cool down slowly?
Sometimes the magma never actually reaches the surface. It rises up, but gets trapped in the Earth's crust. Surrounded by rock, the magma cools down slowly So large crystals are formed. Granite was formed like this.

Other magma escapes from the crust.
For example, most of the rock under the oceans is basalt. Magma comes out of cracks in the sea-bed.
The cold water cools the magma quickly.
That's why basalt has small crystals.

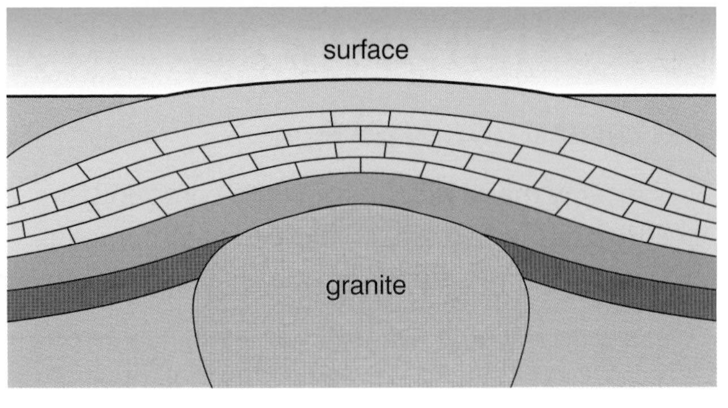

The granite cools slowly.
Can you see where metamorphic rock might be formed in the diagram?

The rock cycle

Look at the rock cycle below:

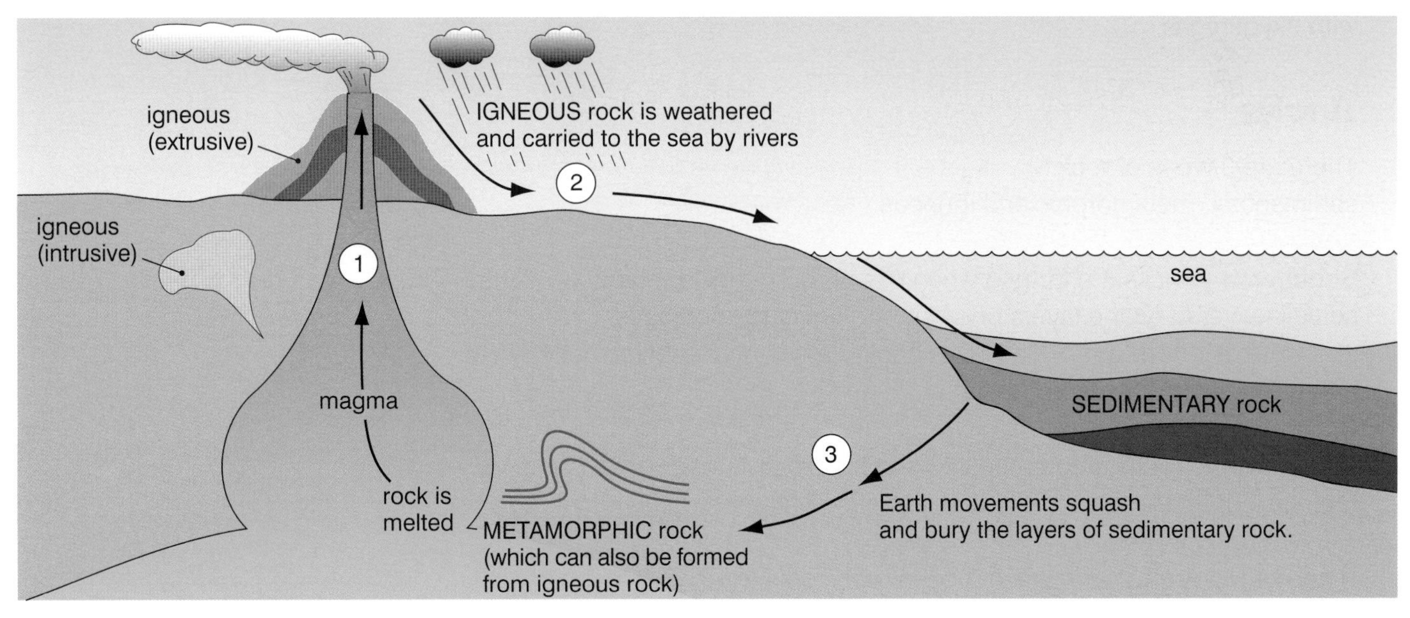

Changes in the rock cycle can take millions of years.

e) Draw a flow diagram to show the changes that take place in the rock cycle.

Remind yourself!

1 Copy and complete:

Igneous rocks form when ...... rock cools down.
They are made up from ......
The size of the ...... depends on how ...... the rock cooled originally. The faster the rock cooled down, the ...... the crystals.
The rock ...... shows how rocks are recycled over time.

2 a) Granite is called an intrusive igneous rock. It cooled down inside the Earth, underground. Basalt formed when lava cools down quickly. It is called an extrusive igneous rock. Which of the two rocks has the larger crystals?

b) Name another igneous rock that has large crystals.

Summary

The atmosphere

The air is made up of about 80% nitrogen gas,
20% oxygen and small amounts of other gases.
These include carbon dioxide, water vapour and noble gases.

In the Earth's first billion years, its early atmosphere came from volcanoes.
It was probably mainly carbon dioxide (like Mars and Venus now).
There was no oxygen. This only arrived once the first plants had evolved.
During photosynthesis, the plants took in carbon dioxide and gave out oxygen.
Most carbon became 'trapped' in fossil fuels and carbonate rocks.

The oceans were formed when water vapour from the volcanoes
fell as rain as the Earth cooled down. The small amount of ammonia
and methane in the early atmosphere was removed when they reacted
with oxygen gas.

Rocks

There are 3 types of rock:
sedimentary, metamorphic and igneous.

Sedimentary rocks are formed when bits of rock, shell or plants
settle in layers. As the layers build up, the pressure increases
and water is driven out. This leaves behind anything dissolved in the water
and the bits are 'cemented' together.

Metamorphic rocks are formed when rocks are put under great pressure
(see next chapter) and/or are heated to high temperatures (without melting).

Igneous rocks are made when molten rock cools down and forms crystals.

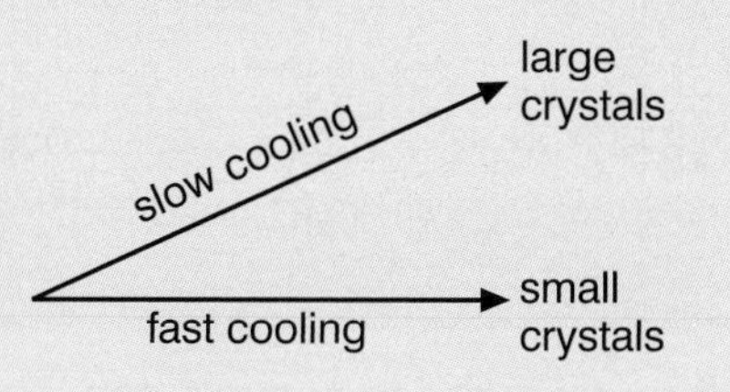

Questions

1 Copy and complete:

Almost four-...... of the air is nitrogen gas, and nearly one-...... is oxygen.

The Earth's early atmosphere came from ......

The ...... formed as the Earth ...... down and water vapour condensed and fell as ......

Most of the early atmosphere was .... gas.

...... gas was made when the first plants developed.

The Earth's crust is made up of 3 types of ......
They are called ......, ... and igneous.

...... rock is made from bits of rock that are deposited in layers.

...... rock is formed when other rock is subjected to very high ...... and/or high ......

Igneous rock was formed from ...... rock that cools down, making ......

The more ...... the rock cools down, the larger the crystals.

2 This question is about the gases in the air. Clean air is a mixture of gases:

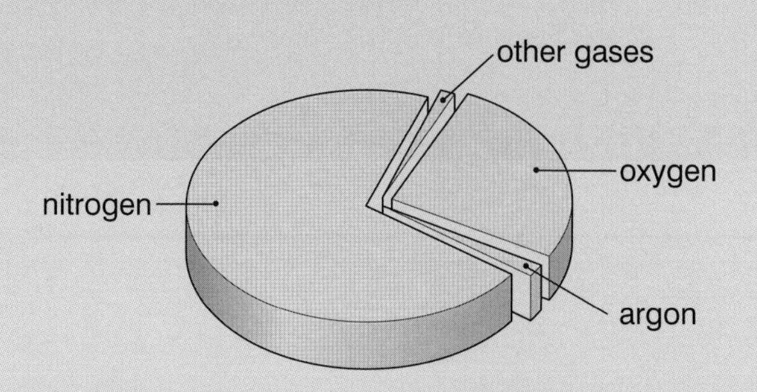

a) Which gas in the air do we need for things to burn?

b) Which gas in the air turns limewater milky (cloudy)?

c) Which gas forms most of the air?

d) Which gas is used inside light bulbs?

e) Name two of the other gases referred to in the pie chart above.

3 The gases in the air have been almost the same now for about 200 million years.

Write a brief history of each of the gases below. Say how they have changed since the early volcanic atmosphere on Earth:

a) carbon dioxide

b) oxygen

c) ammonia

d) water vapour.

4 We sometimes find fossils in rocks.

a) Which type of rock (igneous, metamorphic or sedimentary) are we most likely to find fossils?

b) How can a fossil help us to find out about the history of the rock it was found in?

c) Occasionally you can find fossils in slate. These are usually distorted and twisted. Explain why this has happened to the fossils.

5 The diagram below shows where different rocks are found:

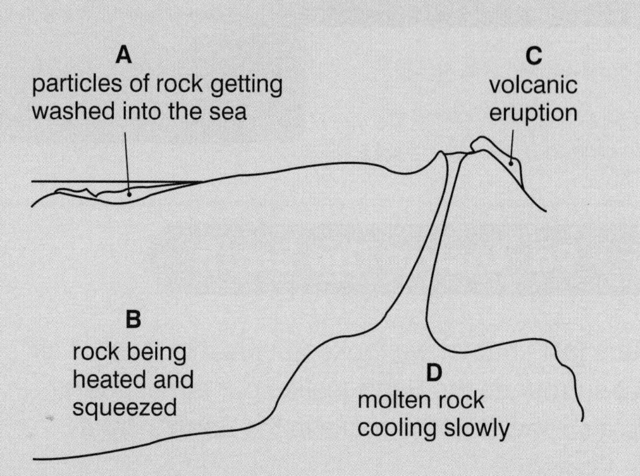

At which place (A, B, C or D) will the rocks in the table below be formed?

Rock	Description and use
slate	This is a hard rock that we can split into thin sheets. We use it to tile roofs. It is a metamorphic rock.
granite	This is a very hard rock used in building. It is an intrusive igneous rock, made from large crystals.
sandstone	This rock is made from rounded grains of sand. It is a sedimentary rock, sometimes used to line the inside of furnaces.
basalt	This very hard rock is an extrusive igneous rock. We use it for making roads.

6 Imagine slicing through a section of the Earth's crust. The diagram below shows the rocks you might find in one place.

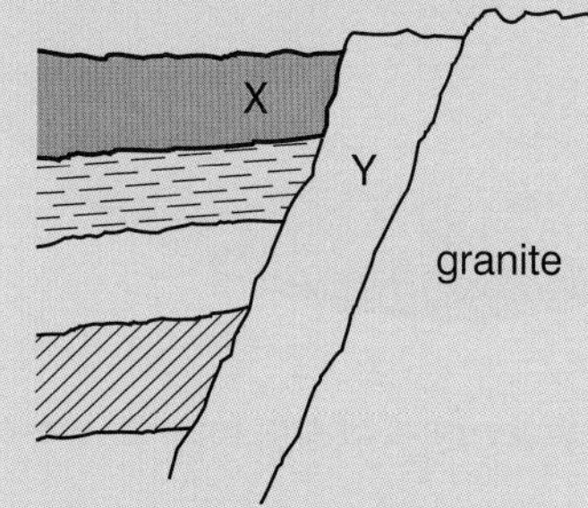

What type of rock would you find at X and Y (igneous, metamorphic or sedimentary)?

Explain how you got your answer.

CHAPTER 9 Structure of the Earth

9a Inside the Earth

In the last chapter we looked at the thin layer of air that surrounds the Earth (called our atmosphere).
We also saw how the rocks in the Earth's crust are formed.
These are very important parts of our planet.
But they are a very small part of it.
Have you ever wondered what's inside the Earth?

Look at the diagram below:

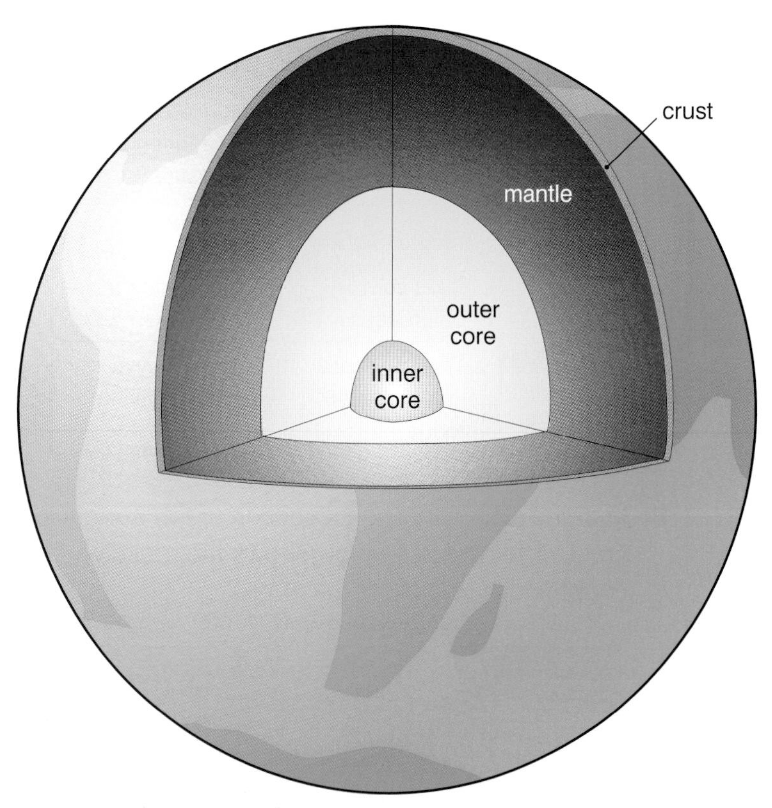

Looking inside our planet.
The Earth's crust and the upper part of the mantle are called its ***lithosphere****.*

Why had people never seen a photo like this before the late 1950s?

a) What is the outer layer of the Earth called?
b) How many layers make up the Earth's core?
c) What do we call the layer above the core?
d) Which parts of the Earth form the **lithosphere**?

The crust

The thin outer layer of the Earth is called its crust.
It's a bit like the shell of an egg.

The core varies in thickness. It can be as thick as 70 km under the continents or as thin as 5 km under the oceans.

Nobody has yet drilled a hole deep enough to pierce the Earth's crust. But we do have evidence of what is down there, for example from volcanoes. Scientists have also calculated the mass of the Earth. From this they have worked out that the materials under the crust must be more dense than the rocks in the crust itself.

Molten rock from under the crust can be studied by vulcanologists.

e) Is the Earth's crust thicker under the continents or under the oceans?

f) How do volcanoes give us evidence of what lies beneath the crust?

The mantle

Underneath the crust we find the mantle.
This goes down almost half way to the centre of the Earth.
Most of the rock in the mantle is hot solid.
But just under the crust and upper part of the mantle
a small amount is almost molten and can flow very slowly.

The core

The core of the Earth is made up of iron and nickel.
Both of these are magnetic metals.
The outer core is liquid (molten metal), whereas
the inner core is solid.

Remind yourself!

1 Copy and complete:

The Earth's thin ...... layer is called its ......
Under this we have the ......

This is mainly solid but can ...... slowly just underneath the ...... and upper part of the ......

The core at the ...... of the Earth is made from ...... and ...... The outer core is ...... but the ...... core is solid.

2 Imagine you are the head of a scientific team. You have invented a new machine that can tunnel through rock, carrying passengers at the same time.

a) Draw a sketch of your new machine.

b) Describe what you would see on the first trip from one side of the Earth to the other – by the shortest route possible!

9b Drifting continents

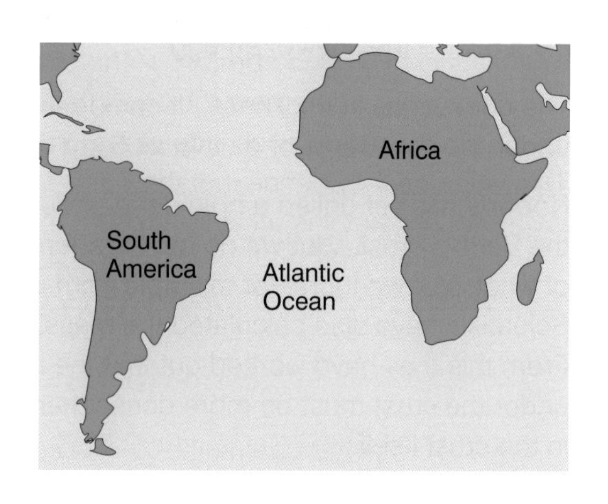

Have you ever noticed that the continents of Africa and South America look like pieces in a jigsaw. When you look at an atlas, you can almost imagine that at one time they were joined together. But surely that's a silly idea. After all, how could continents move thousands of miles apart? This is what scientists were arguing about for much of the last century.

In 1915 a German scientist called **Alfred Wegener** put forward his theory of 'drifting continents'. People had already commented on the strange matching shapes of Africa and South America. But Alfred went one daring step further. He said that they were once joined and had drifted apart. He found rocks and fossils on both continents that matched up. Look at the diagram below:

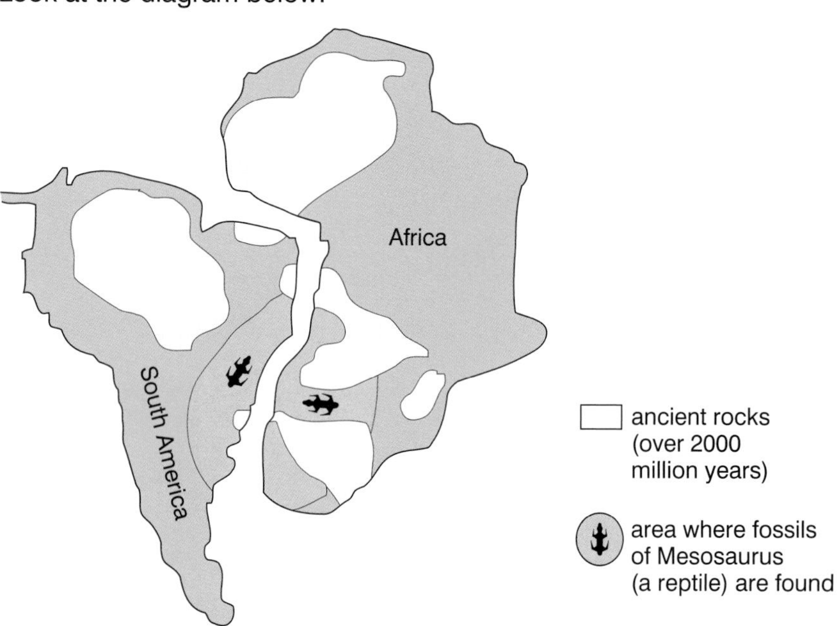

The shapes of South America and Africa slot together as in a jig-saw. This gives us evidence that the continents were once joined and must have drifted apart.

But other scientists didn't think much of Alfred's theory. Many believed that mountains and oceans were formed as the Earth first cooled down and formed its crust. The crust would take up less space than molten rock so it would shrink and wrinkle up as it formed. They had a theory that made sense to them. They just didn't believe that continents could move – and Alfred couldn't explain how they could move. So most scientists chose to stick with their old theory.

a) Why didn't other scientists accept Alfred Wegener's theory?

It wasn't until after Alfred had died that his theory was accepted. Other people noticed matching rocks and fossils on other pairs of continents. Then eventually in the 1960's scientists exploring the bottom of the ocean noticed cracks where new rock was being made. The Earth's crust on each side of the crack was moving apart. This was direct evidence that the Earth's crust is still moving – very slowly, but it is moving. So continents can drift apart!

b) Which piece of evidence finally convinced scientists that Alfred Wegener's theory was a good one?

Today most scientists think that all the continents were once joined together. They have called the huge area of land (or land mass) Pangaea.
Look at the diagram below:

Can you see the shapes of any of today's continents outlined on Pangaea?

There was just one thing left to explain now.
What causes the movement of the Earth's crust.
This is covered on the next pages.

Alfred said that all continents drifted.
'Yes, the land masses actually shifted!'
'That sounds quite weird!'
Other scientists sneered.
But now his idea's been accepted.

Remind yourself!

1 Copy and complete:

Alfred ...... suggested his theory of ...... continents in 1915. He used evidence from rocks and ...... found in both ...... and South America. He said that they were once ...... but had ...... apart.

However, it wasn't until about 50 years after that his ideas were ......

2 a) Why was Alfred Wegener's theory described as daring.

b) Imagine you are one of Alfred's fellow scientists at the time.

Write a letter to Alfred explaining why you do or don't support his theory.

c) Find out about the life of Alfred Wegener.

9c Moving plates

We now believe that the Earth's crust and the top of its mantle are made up from huge slabs of rock called **tectonic plates**. This is called the Earth's **lithosphere.**

You can work out where the edges of the plates are by looking at where we get earthquakes and volcanoes. These are shown on the map below:

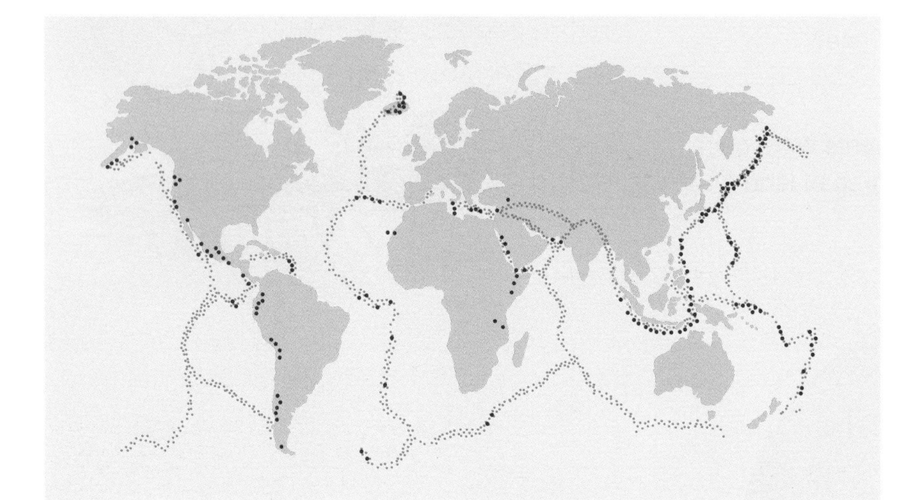

Key:
 earthquake zones
volcanoes

a) Is Britain likely to suffer from earthquakes or volcanoes?

b) Name 3 countries where you would expect earthquakes.

The main plate boundaries are shown below:

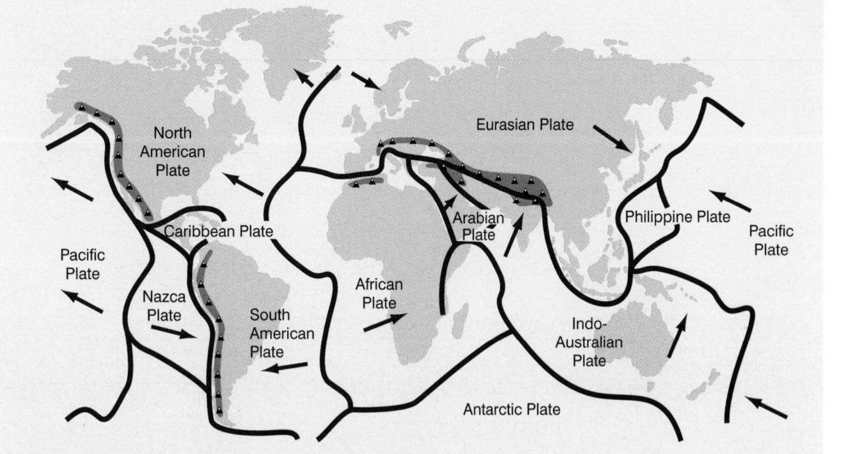

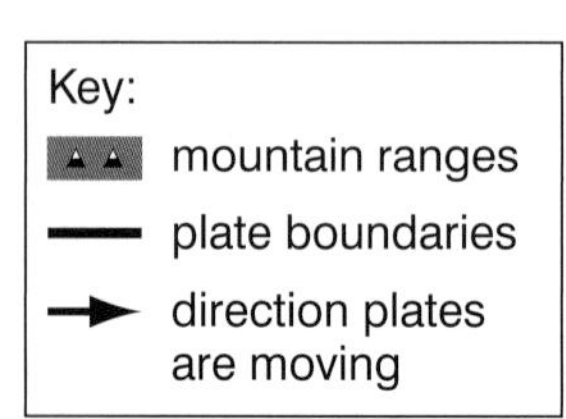

c) Explain your answer to question a) above.

d) What else, besides earthquakes and volcanoes, can you find at some plate boundaries.

e) Name 2 plates that are moving away from each other.

f) Name 2 plates that are moving towards each other.

Volcanoes

You can imagine the magma (molten rock) rising through the crust to the surface where two plates meet. It erupts from the surface as lava.
Over time, as the lava cools and turns solid, a volcano forms. Volcanoes can have different shapes depending on the thickness of the lava. Look at the diagrams below:

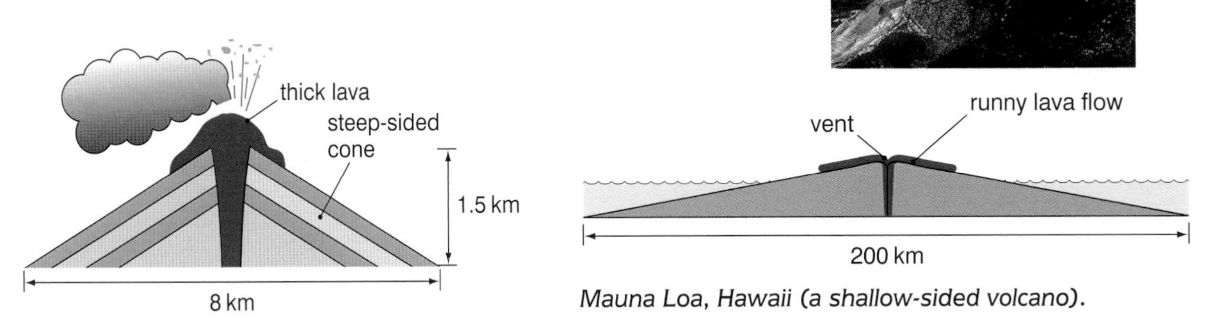

Mauna Loa, Hawaii (a shallow-sided volcano).

Mount Pelée, West Indies (a steep-sided volcano).

Earthquakes

The map at the bottom of the last page shows the direction that the plates are moving. Here is one theory to explain how they move:

Radioactive rocks produce heat, deep in the Earth's mantle. The hot rock in the mantle very slowly rises, then falls as it cools. This sets up **convection currents** under the plates and causes them to move slowly (at about the rate your finger-nails grow!). Look at the diagram opposite:

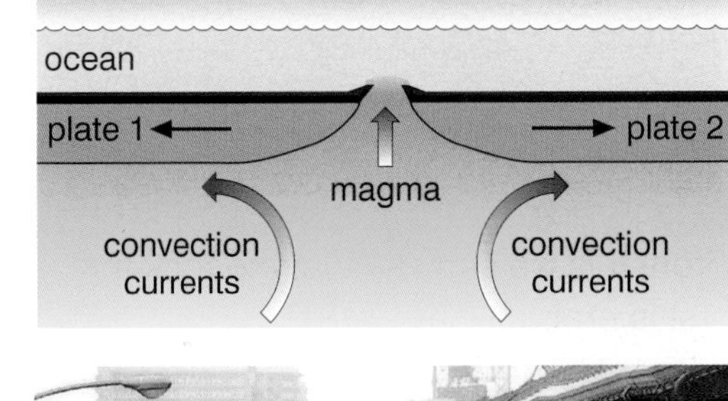

The plates can sometimes slip past each other. Where this happens suddenly, we get an earthquake. The edges of the plate are jagged and the force builds up until the plates slip.
Unfortunately, scientists can't predict yet when this will happen. However, they are getting better at looking for warning signs using sensitive equipment.

This earthquake in Kobi killed 4,571 people

1 Copy and complete:

The Earth's crust and top of its mantle are made up of giant ...... plates.

Radioactive rocks give out ...... which causes ...... currents under the plates.

This makes them move, sometimes ...... past each other and causing ......

Where magma breaks through the Earth's ...... we get ......

2 a) Why are some volcanoes low and wide whereas other are higher and steeper?

b) What could affect the properties of the lava given out from a volcano?

c) Find out about a recent earthquake or volcanic eruption and write a newspaper article about it.

▶▶▶ 9d Clues in the rock

In the last chapter we looked at the 3 different types of rock and how they were formed.
We can also get clues about what happened in the past from the pattern of rocks in the crust.
It's a bit like being a detective. The clues are in the rock.
You can use them to work out what happened long before dinosaurs lived on Earth.

For example, look at the layers of sedimentary rock below:

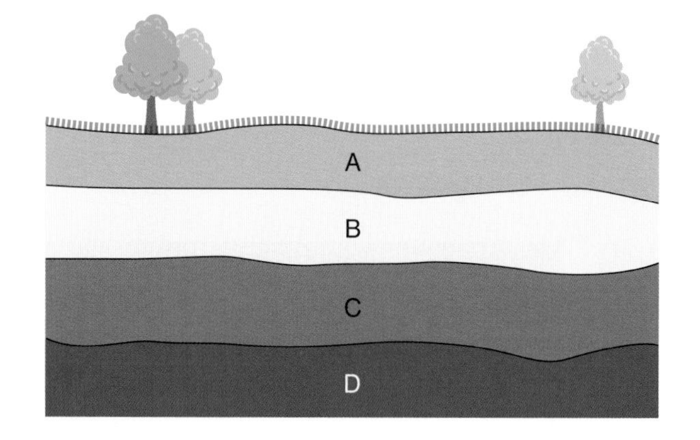

From this we can say that the rock that was formed first is probably layer D.
If layer B is rock salt, we can guess that there was once an inland sea here that evaporated and left behind the salt.

a) Which is probably the youngest rock layer in the diagram above?

b) What might have happened to make the inland sea dry up?

Layers of rock are often **folded** showing that they have been put under great pressure. Look at the photo opposite:
The folding was probably caused by plates colliding together as they form mountains. Metamorphic rock is made here.
Sometimes the layers actually snap. This forms a **fault** in the rock.
Look at the diagram below:

Folds in the layers of rock.

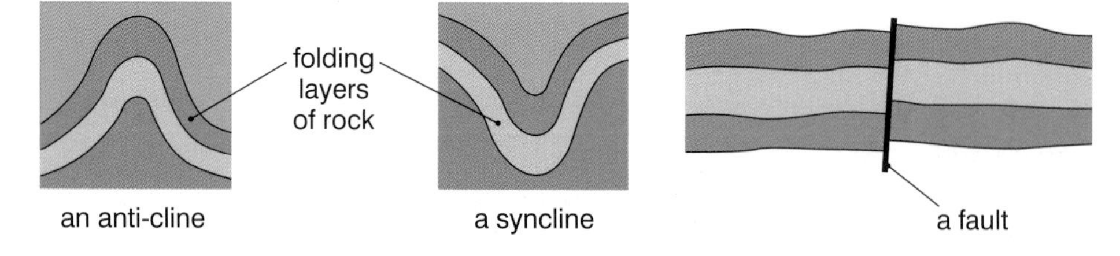

c) Which of the 3 types of rock are formed when plates collide?

d) Draw a fold in some layers of rock and draw arrows to show the forces applied to make the fold.

Remember that metamorphic rock is also formed when rock is heated to high temperatures without melting. For example, in the diagram below we can work out the order that events happened:

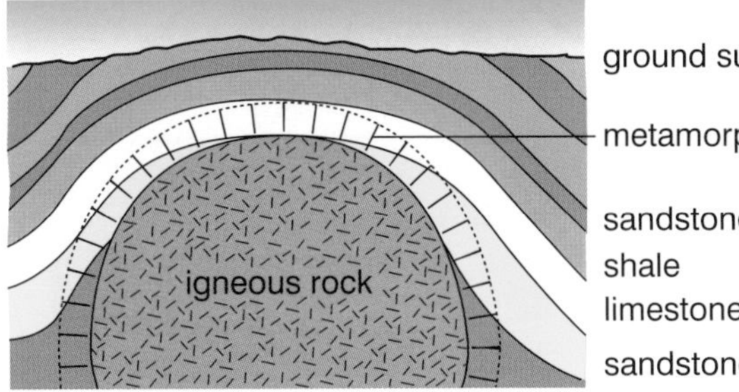

First of all the layers of sedimentary rock were laid down.
Then magma rose towards the surface.
This caused the sedimentary rock near it to be 'baked'.
This turned into metamorphic rock.
Gradually the magma cooled down and formed igneous rock.

e) Which is the oldest rock in the diagram above?

Rocks are slowly worn away over time by erosion. So today you might find the rock structure above like this:

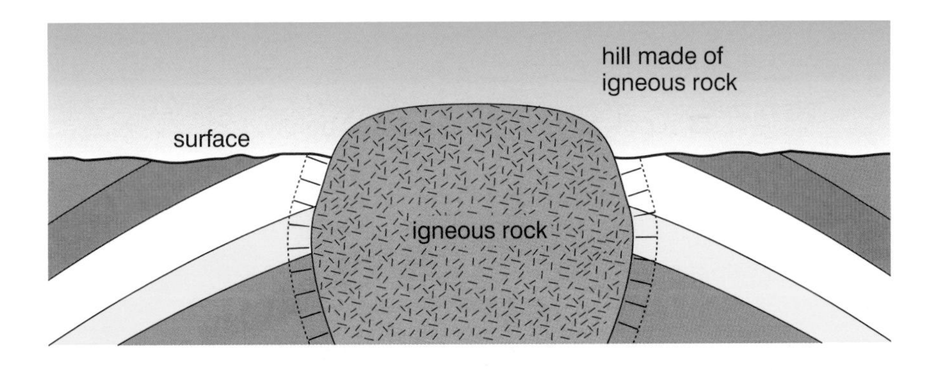

Remind yourself!

1 Copy and complete:

We can get clues about what happened millions of years ago from ......

For example, younger ...... rock usually lies on top of older rock.

The movement of the Earth's plates can form ...... rock under great pressure.

Folds and ...... can show the direction of the forces applied.

2

a) This rock was found in a mountain range. Is it igneous, metamorphic or sedimentary?

b) Describe how the rock was formed.

Summary

The Earth is made up from:

- a thin outer **crust**,
- a **mantle** (under the crust stretching almost half way to the centre of the Earth),
- a **core** (the outer core is liquid; the inner core is solid; and both parts are made from iron and nickel).

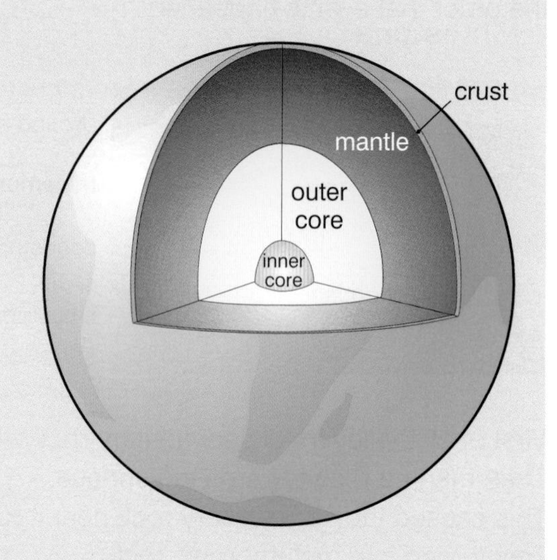

The Earth's crust and uppermost part of its mantle (called the **lithosphere**) is split up into **tectonic plates**.
These move very slowly on convection currents set up in the mantle. One theory suggests that the heat comes from radioactive rocks.

We can work out the plate boundaries by looking at where we get earthquakes and volcanoes.

Where the plates collide, mountains are built (replacing those worn away over millions of years by weathering and erosion).
You find metamorphic rocks in these mountain ranges.
This is due to the great pressure and high temperatures produced as mountains form.

Questions

1 Copy and complete:

The Earth has an outer ……, followed by a ……, then an …… liquid core, and a …… inner core at its centre.

The outer shell of the Earth is split into …… plates.

Heat produced by …… rocks in the mantle causes convection …… that move the plates slowly. The movement of the plates can cause …… We can use these, plus volcanoes, to mark the …… of the plates on a map.

2 a) Draw a labelled diagram to show what the Earth would look like if it was sliced in half.

b) Which part of your diagram causes the Earth to behave like a giant magnet?

c) Which part of the Earth is called the lithosphere?

3 Look at the rock structures below:

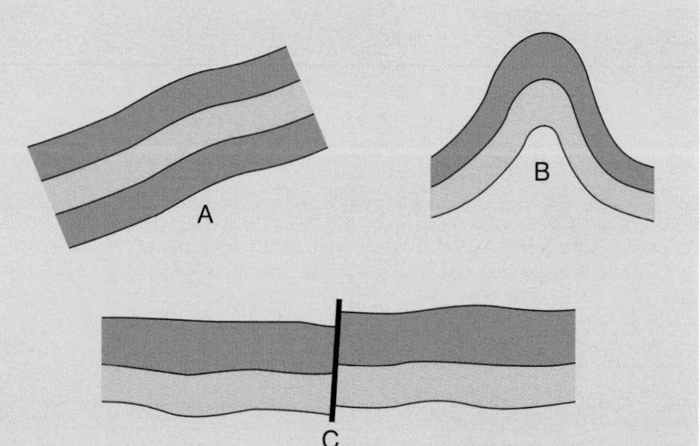

a) Use the words below to describe the structures labelled A, B and C:

tilted   folded   fault

b) What causes the structures in the rocks above?

▶ Limestone

1 The flow chart shows the action of heat on calcium carbonate.

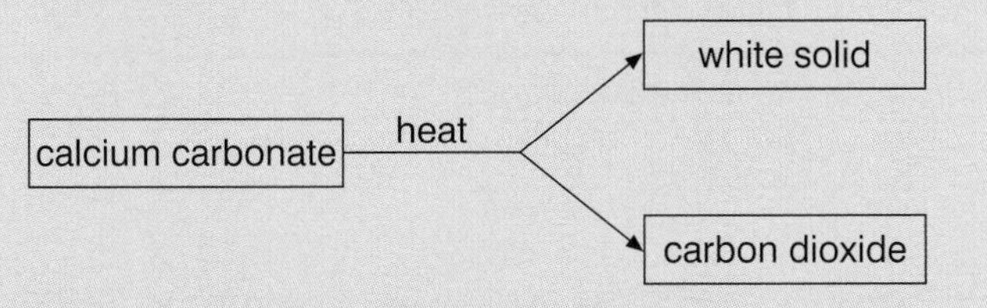

(a) Write a word equation for the action of heat on calcium carbonate. (1)

(b) Give a test for carbon dioxide and its result. (2)

(c) A few drops of water were added to the white solid. Some of the water turned to steam.

(i) Explain what caused some of the water to turn to steam. (2)

(ii) Name the substance formed when water was added to the white solid. (1)

(AQA SEG 1998)

2 **(a)** The flow diagram below shows some of the substances which can be made from limestone.

Complete Boxes 1 and 2 below using words from the list.

calcium chloride	**calcium hydroxide**	**calcium sulphate**
marble	**quicklime**	**sandstone**

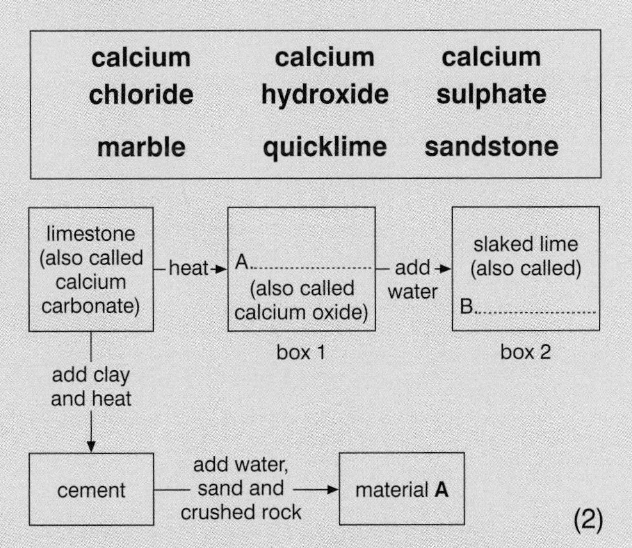

(2)

(b) When cement is mixed with water, sand and crushed rock, a chemical reaction takes place which produces material **A**. Name material **A**. (1)

(AQA 2000)

3 Limestone is an important raw material. Some of its uses and products are shown in the diagram. Some are missing.

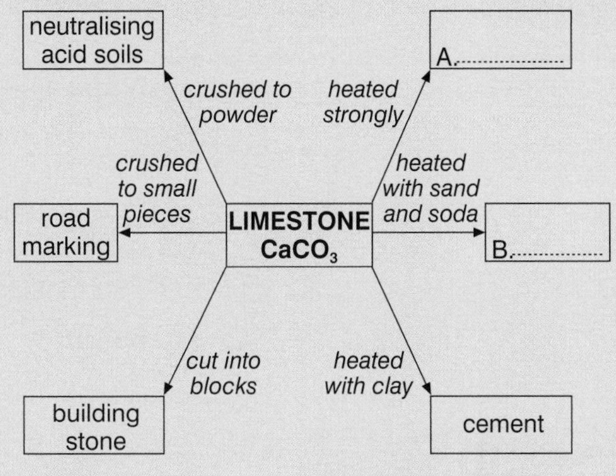

(a) What are the names of missing products A and B? (2)

(b) Bags of cement powder are labelled with this hazard symbol.

(i) What does this hazard symbol tell you about cement powder? (1)

(ii) Give **one** precaution you should take when handling cement powder. (See page 123) (1)

(AQA SEG 2001)

4 Portland cement was invented by Joseph Aspdin, a builder from Leeds. The flow diagram shows how cement is made.

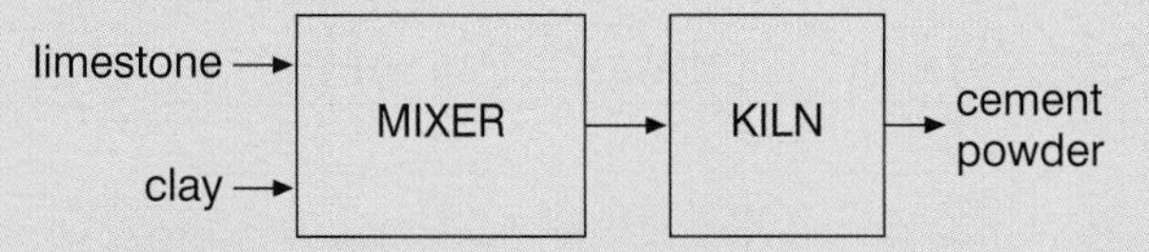

(a) What are the **two** raw materials used to make cement? (1)

(b) Cement is mixed with three substances to make concrete.

Choose from the box the **three** substances used. (3)

crushed rock	**iron ore**	**quicklime**	
sand	**slag**	**soda**	**water**

(AQA SEG 1999)

▶ Oil

5 The diagram shows some of the processes in an oil refinery. Some of the substances in the oil refinery are represented by the letters **A** to **J**.

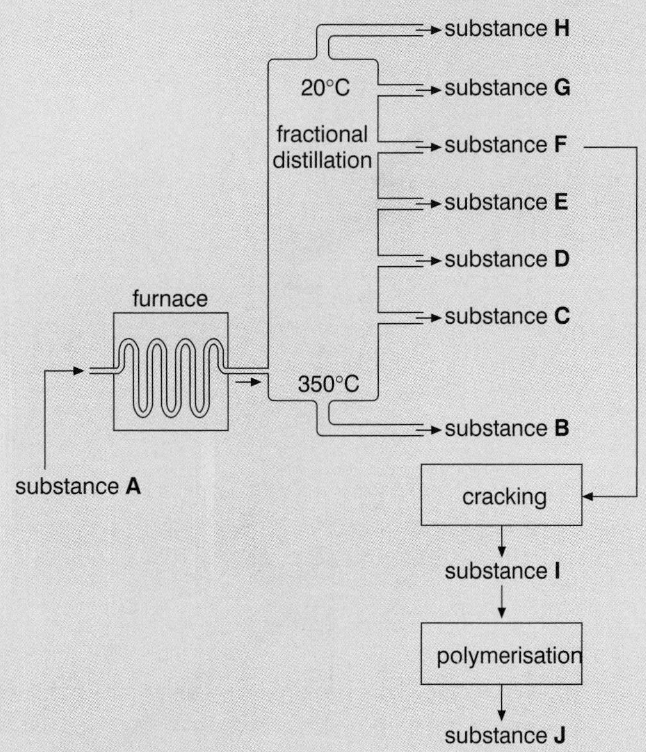

Six of these substances are described in the table. Match each description to the correct letter, **A** to **J**. One letter has been done for you.

Substance (A–J)	Description of substance
	(i) crude oil
	(ii) petroleum gases (the fraction with the lowest boiling point)
G	petrol
	(iii) polymer
	(iv) small useful molecules made by the breakdown of larger molecules
	(v) road tar (the fraction with the highest boiling point)

(5)

(AQA SEG 2001)

6 The high demand for petrol (octane) can be met by breaking down longer hydrocarbons, such as decane, by a process known as cracking.

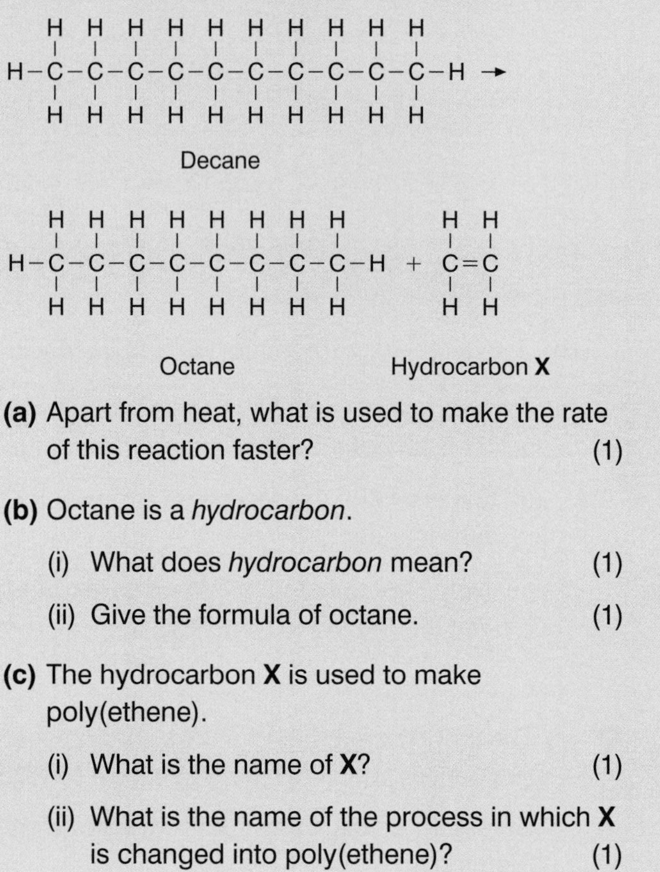

(a) Apart from heat, what is used to make the rate of this reaction faster? (1)

(b) Octane is a *hydrocarbon*.

(i) What does *hydrocarbon* mean? (1)

(ii) Give the formula of octane. (1)

(c) The hydrocarbon **X** is used to make poly(ethene).

(i) What is the name of **X**? (1)

(ii) What is the name of the process in which **X** is changed into poly(ethene)? (1)

(AQA SEG 2000)

7 Crude oil is a fossil fuel.

(a) Complete the sentence by choosing the correct words from the box.

energy	long	plants	rocks	short

You may use each word once or not at all.

Fossil fuels were formed from the remains of …… and animals over a …… period of time. (2)

(b) Give **one** other example of a fossil fuel. (1)

(c) In the future there may be no fossil fuels left. Why? (1)

(AQA SEG 1999)

8 The gas used as a fuel for heating in most homes is methane, $\mathbf{CH_4}$.

(a) It is very important to have a good air supply when methane burns. Explain why. (2)

(b) The word equation when methane burns in a good air supply is:

methane + oxygen → carbon dioxide + water

(i) Copy and balance the chemical equation for this reaction.

$$\ldots\ CH_4(g) + \ldots\ O_2(g) \rightarrow \ldots\ CO_2(g) + \ldots\ H_2O(g)$$

(1)

(c) The experiment shown was used to test the gases formed when methane burns in a good air supply.

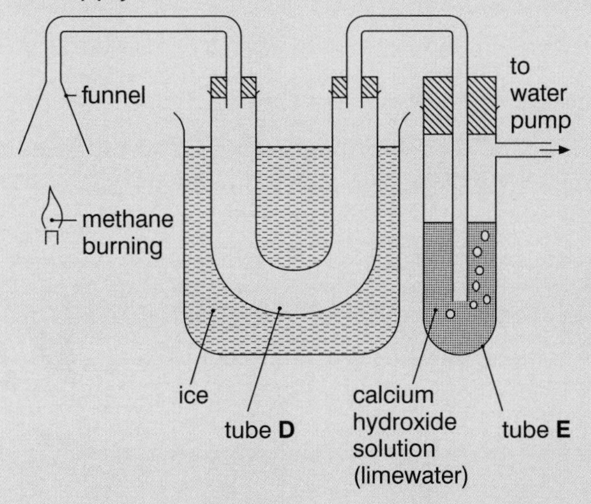

(i) Explain why the water formed collects in tube **D**. (2)

(ii) Give a chemical test for water and its result. (2)

(iii) The reaction that happens in tube **E** is:

$$Ca(OH)_2(aq) + CO_2(g) \rightarrow CaCO_3(s) + H_2O(l)$$

Describe and explain the change you would see in tube **E**. (2)

(AQA SEG 1999)

9 Since 1850 there has been an increase in the amount of carbon dioxide in the atmosphere.

The table shows the estimated amounts of carbon dioxide in the atmosphere.

Year	Carbon dioxide in parts per million
1850	270
1900	285
1960	315
1980	335
2030	600

(a) (i) Plot the points on a graph. (2)

(ii) Finish the graph by drawing the best curve through the points. (1)

(iii) Use your graph to find the estimated amount of carbon dioxide in the atmosphere, in the year 2000.

Your answer will be in parts per million (1)

The increase of carbon dioxide in the atmosphere has led to the Greenhouse Effect.

The diagram shows the Greenhouse Effect.

(b) Use the diagram to help you answer the questions.

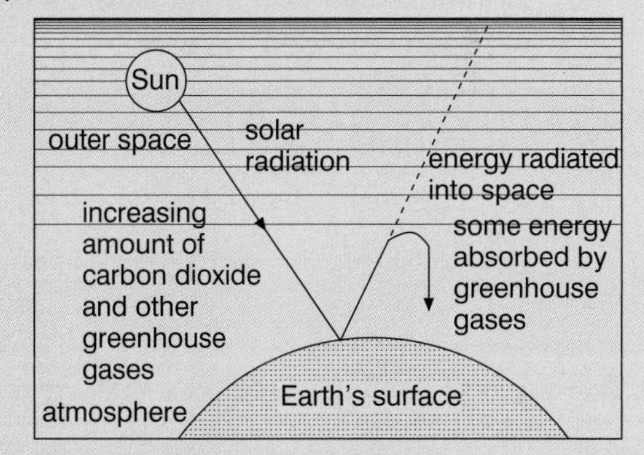

Here are some sentences about the Greenhouse gases.

Finish the sentences by selecting the **best** word.

Solar radiation passes through the (i) …… and warms the Earth's surface. The Earth radiates (ii) …… back into space. Some energy is absorbed by the (iii) …… ……. This causes the air to become (iv) ……. (4)

(c) (i) Explain how human activity is increasing global warming. (2)

(ii) Describe **two** effects global warming will have on the environment. (2)

(OCR Suffolk 1999)

Further questions on the Earth and its resources

▶ Air and rocks

10 **(a)** The Earth's atmosphere is a mixture of gases.

Use words from the list to complete the table about the composition of the atmosphere.

carbon dioxide nitrogen noble gases oxygen water vapour

Gases	Proportion of gases in the Earth's atmosphere today
1	about four-fifths ($^4/_5$)
2	about one-fifth ($^1/_5$)
3	very small
4	
5	

(3)

(AQA SEG 2000)

11 **(a)** Millions of years ago, the atmosphere contained:

ammonia carbon monoxide methane nitrogen oxygen steam

The amounts of these gases have changed over millions of years.

(i) State ONE gas which has decreased. (1)

(ii) State ONE gas which has increased. (1)

(EDEXCEL)

12 Sand and dead sea creatures form layers on the seabed.

(a) What name is given to the material which forms these layers?

A crystals **B** plates **C** salt **D** sediment

(1)

(b) Rocks are classified into three main types

igneous metamorphic sedimentary

(i) When the layers on the seabed are buried they turn into rock.
What type of rock is formed? (1)

(ii) High temperature and pressure can change rock without melting it.
What type of rock is formed? (1)

(c) High temperature and pressure can change limestone.

What is the name of the rock formed when limestone is changed in this way? (1)

(EDEXCEL 1999)

▶ Structure of the Earth

13 The sketch below was made on a field trip in Wales.

A, **B** and **C** are three different types of rock.

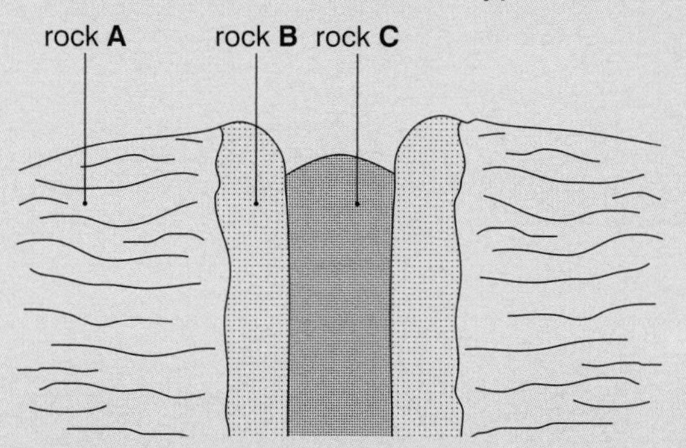

The diagrams below show what the three rock types look like under a microscope.

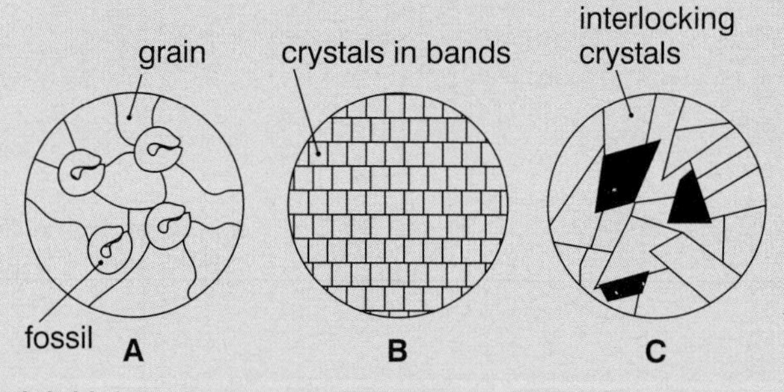

(a) Using the words in the box below copy and complete the table that follows:

granite igneous limestone marble metamorphic sedimentary

Rock	*Type of rock*	*Name of rock*
A		
B		
C		

(3)

(b) Place the rocks, **A**, **B** and **C** in order of their age (youngest first).

(WJEC)

Section Three

Chemical reactions and their useful products

In this section you will find out more about chemical reactions and how they are useful to us.
You will learn about the factors that affect rates of reaction, the energy changes in reactions and how we use reactions to make fertilisers.

CHAPTER 10

RATES OF REACTION

10a Measuring rates of reaction

Some reactions are fast and others are slow.
When dynamite explodes, the reaction is very fast.
But then so is the reaction between an acid and an alkali.
An example of a slow reaction is iron rusting.

a) Try to think of another example of a fast reaction and another slow reaction.

Whether you are frying an egg or managing a chemical factory, it is important to know how quickly reactions go.
We call the speed at which a reaction goes its **rate of reaction**.

We can't work out the rate of a reaction from its equation.
You have to do an experiment to find out.
You might choose to look at how quickly a reactant is used up.
On the other hand, you could measure how quickly a product of the reaction is formed.

Here are two ways that we can measure the rate of a reaction if a gas is given off:

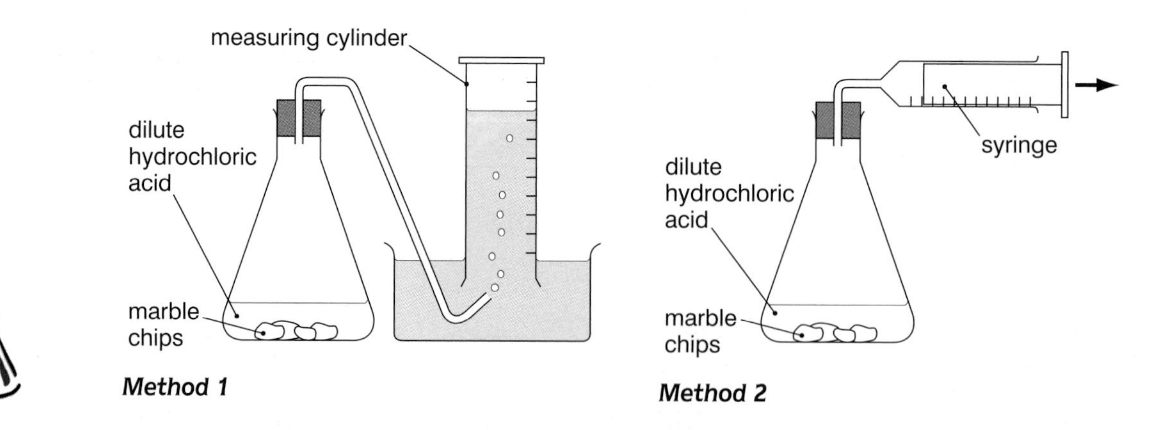

b) What do we use to measure the volume of gas in Method 1?

c) What do we use to measure the volume of gas in Method 2?

d) What else would you need to measure the ***rate*** at which gas is given off?

e) Draw the top of the table you would use to show your results. Show the headings clearly.

You can also measure the mass of gas given off as time passes:

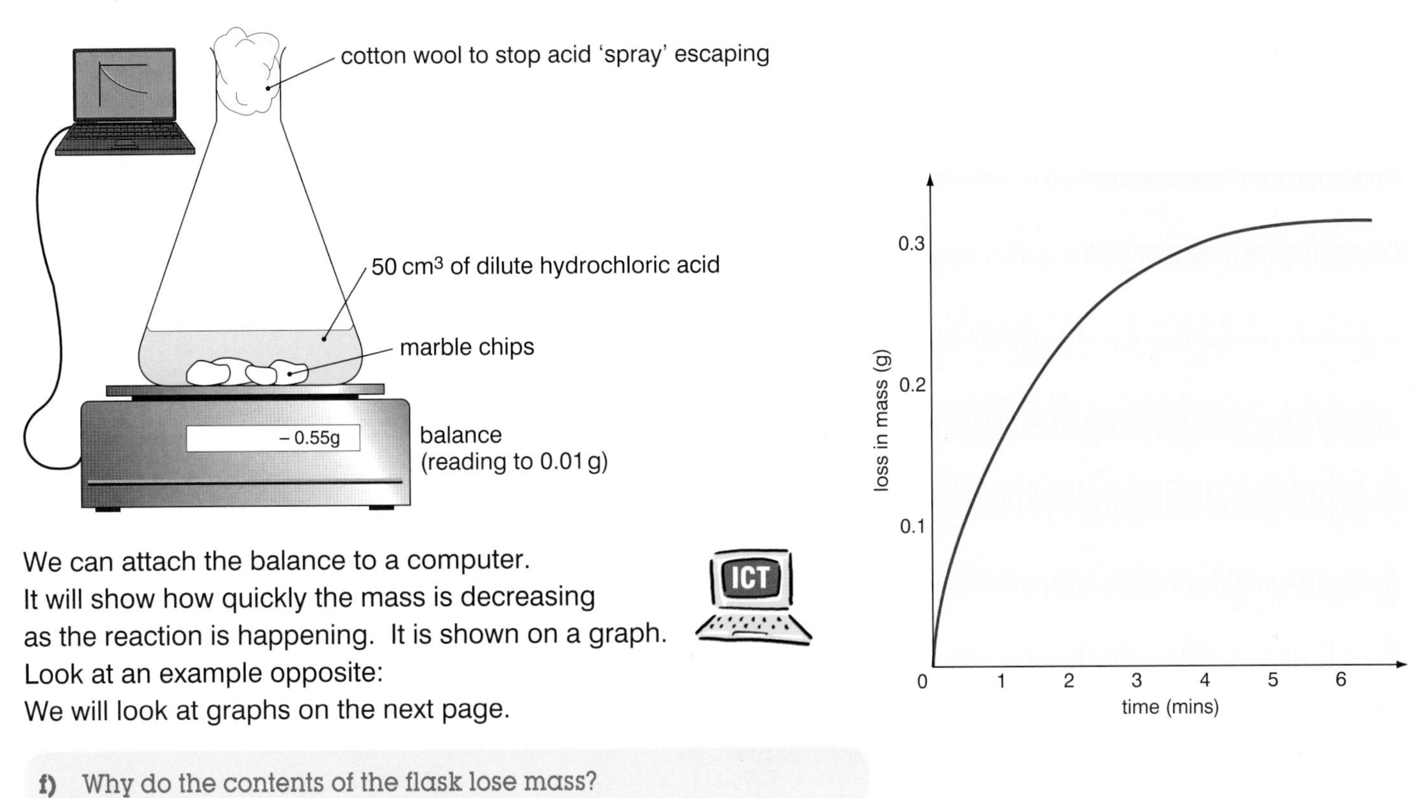

We can attach the balance to a computer.
It will show how quickly the mass is decreasing as the reaction is happening. It is shown on a graph.
Look at an example opposite:
We will look at graphs on the next page.

f) Why do the contents of the flask lose mass?

When planning experiments in this topic it is important to know these hazard symbols:

Oxidising
These substances provide oxygen which allows other materials to burn more fiercely.

Toxic
These substances can cause death. They may have their effects when swallowed or breathed in or absorbed through the skin.

Corrosive
These substances attack and destroy living tissues, including eyes and skin.

Highly flammable
These substances easily catch fire.

Harmful
These substances are similar to toxic substances but less dangerous.

Irritant
These substances are not corrosive but can cause reddening or blistering of the skin.

Remind yourself!

1 Copy and complete:

The rate of a reaction tells us how …… a reaction goes.

We have to carry out an …… to find out the rate of a reaction.

2 When magnesium reacts with dilute sulphuric acid, hydrogen gas is given off.

a) Draw the apparatus you could use to measure the rate of this reaction.

b) Explain how your experiment works.

10b Looking at graphs

Every graph 'tells a story'.
And you need to be able to interpret graphs.
We can use them to tell us about rates of reaction.

Let's look at an example.
We can use results from an experiment measuring the volume of gas given off.

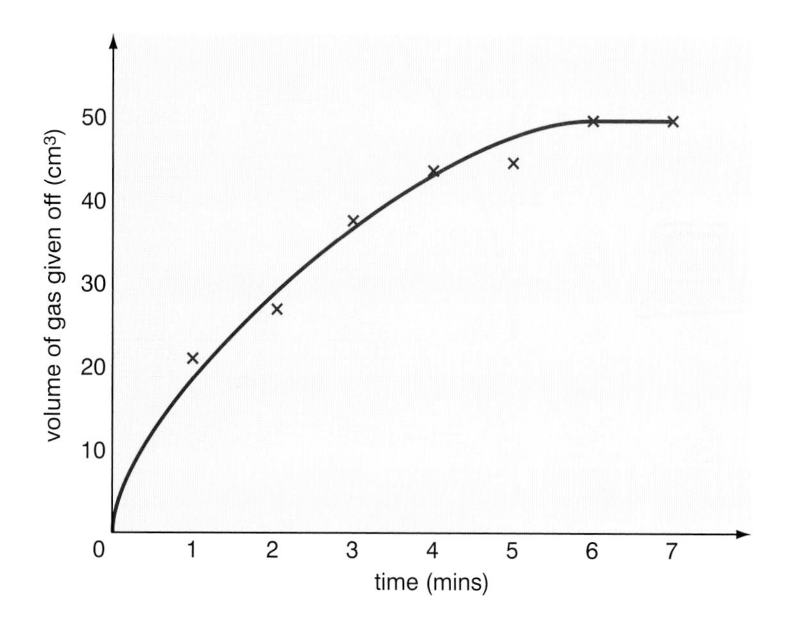

Notice that the graph is a smooth curve.
You don't join the points 'dot-to-dot' with a ruler.

a) How much gas was given off in the first minute?
b) How much gas was given off between the 4th minute and the 5th minute?
c) Was the reaction faster at the start or at the end?
d) How can you tell when the reaction was finished?

The steepness of the ***slope*** on your line graph tells you the rate of reaction at any time.

The steeper the slope, the faster the reaction.

Look at the graph opposite:
It shows what you can find out about a reaction by using a graph.

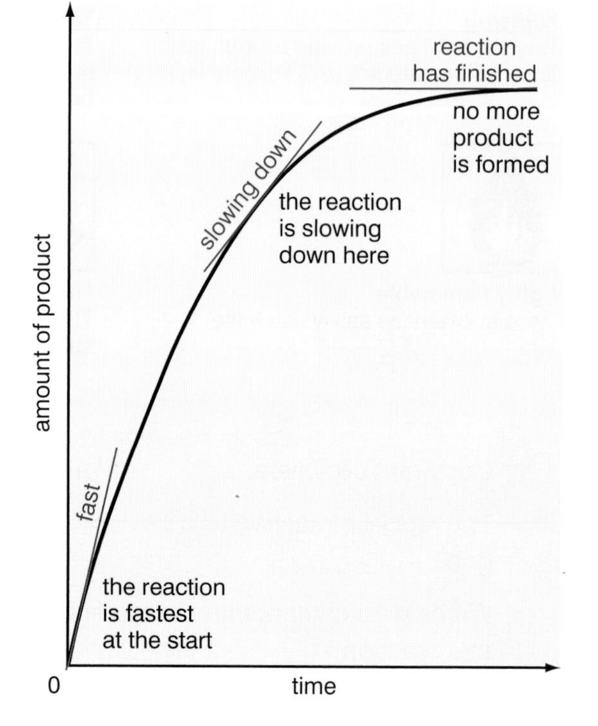

We can compare two reactions on a graph:

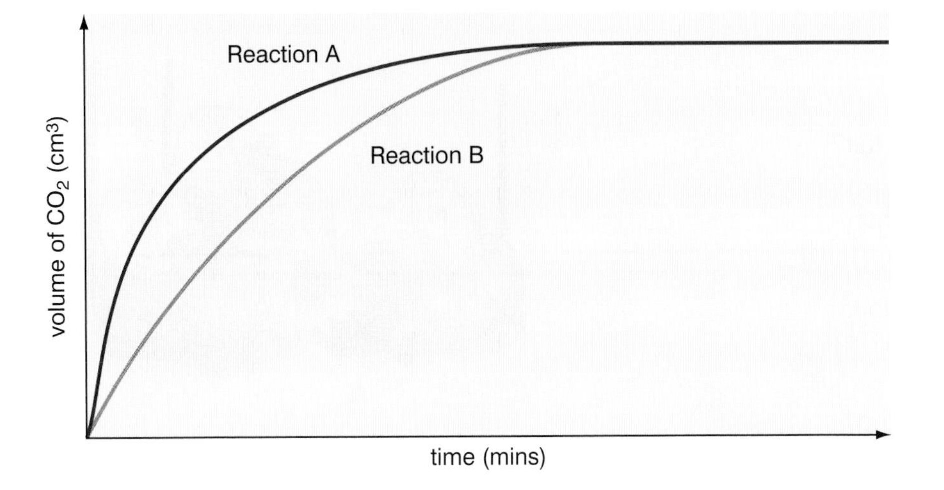

Remember that the slope tells us how quickly the reaction is going.

e) Which is the faster reaction, A or B?

You might also see some graphs where the loss in a reactant is used to follow the rate of reaction. This type of graph looks like this:

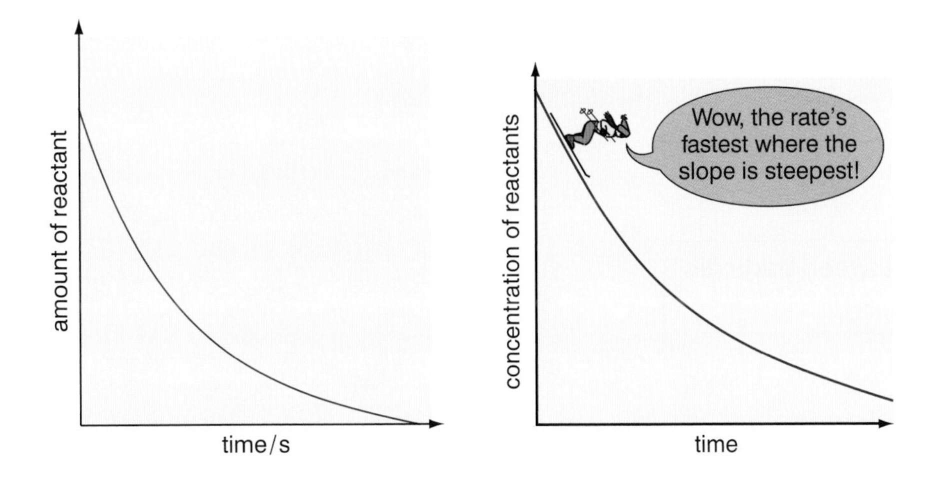

Remind yourself!

1 Copy and complete:

We often follow what happens in a reaction by drawing a ……

The …… the slope of the line, the faster the reaction.

When a line levels out the reaction has ……

2 Look at the graph just above 'Remind yourself!':

a) Why does it slope downwards?

b) What has happened when the line hits the bottom axis?

c) Sketch a graph to show the line if the amount of products had been measured.

10c Colliding particles

Do you enjoy going on the dodgem cars at a fair?
The dodgem cars often crash together.
This is what happens as particles react together.
They ***collide*** with each other.

If the collision is hard enough, we get a reaction.
We call this an **effective collision**.
If the particles only bump together gently,
they just bounce off each other without reacting.

When a reaction does happen, the particles are re-arranged and change into a new substance (unlike a dodgem car!).

This is called the **collision theory**.

Particles must collide, with enough energy, before they can react.

In this topic we are looking at ***rates*** of reaction.

a) What kind of things can you do to make a reaction happen faster?

b) What happens to the number of effective collisions in a given time if a reaction speeds up?

According to the collision theory:

The more effective collisions there are between particles in a given time, the faster the reaction.

We can use the collision theory to explain the factors that affect rates of reaction.
In the next few pages we will look at each of these factors.
They are listed below:

Rates of reaction are affected by:
- **surface area**
- **concentration**
- **temperature**
- **catalysts**

c) Look up the word catalyst. What is a catalyst?

Effect of surface area

Have you ever tried boiling potatoes?
Have you noticed that the larger ones are sometimes hard?
The large ones always take longer to cook than the small ones.
Or how about lighting a fire? It's much easier to start a fire with wood shavings than with blocks of wood.

A potato chopped into small bits has a ***larger surface area*** than the whole potato.

d) Which has the larger surface area – a block of wood or the same block shaved into thin strips?

The smaller the pieces of a solid, the larger its surface area.

Reactions with solids are faster if the solid has a large surface area.

Imagine a solid cut up into little pieces.
Its reactions will be faster because more of its particles are open to attack.
For example, an iron nail reacts much more slowly when you heat it than iron filings.
Look at the diagram below:

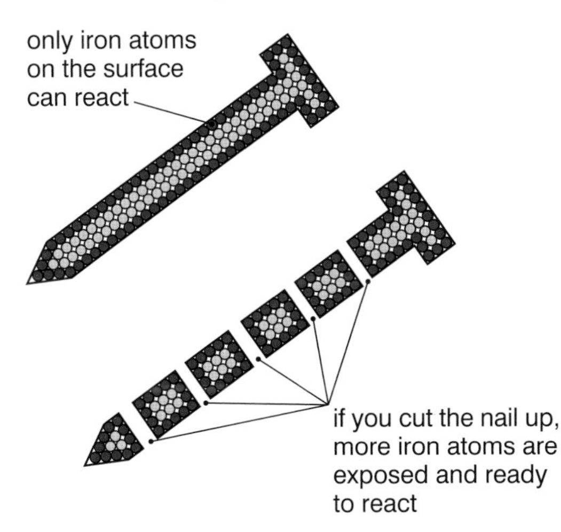

When a solid lump is cut in to pieces,
Its rate of reaction always increases.
In the lump, most particles are locked up inside,
In order to react, they have to collide!
But powders have lots of particles exposed,
If cut fine enough, they might even explode!

Remind yourself!

1 Copy and complete:

Particles must …… before they can react.
They must have enough …… when they hit each other to cause a ……

The more often they …… the faster the reaction.

In solids, the greater the …… area, the faster the reaction.

2 Zinc reacts with dilute acid, giving off hydrogen gas.

a) You can get zinc as strips of metal, granulated zinc (small lumps) or as zinc powder. Which form of zinc will react fastest with the acid?

b) Explain your answer to a) using the collision theory.

▸▸▸ 10d Effect of concentration

Do you like your orange squash strong or weak?
Look at the 3 glasses opposite:
They each contain a solution of orange squash.

We have the same volume of drink in each glass.
However, glass A has most orange squash in it.
Glass C has less orange squash and more water.
So the orange squash is ***more concentrated*** in glass A.

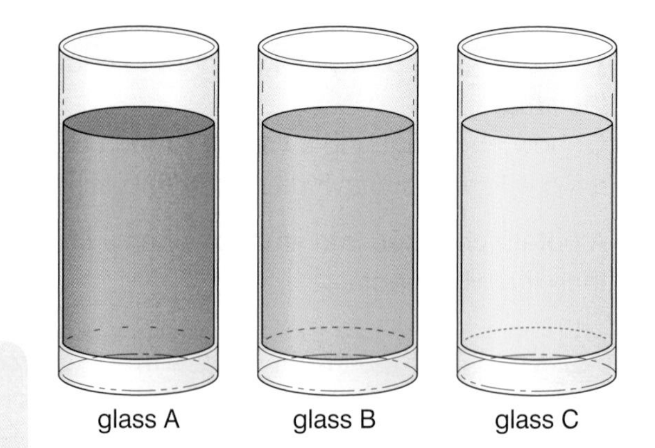

a) List some other solutions we use as concentrated or more dilute solutions.

The orange in glass A tastes stronger because it has
more particles of squash in the same volume of water.

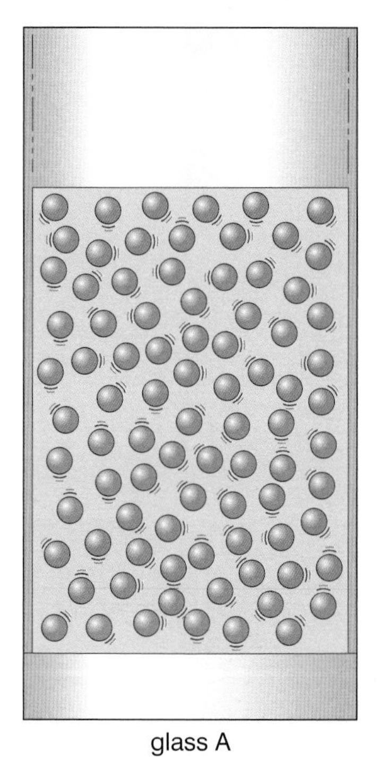

glass A

glass B

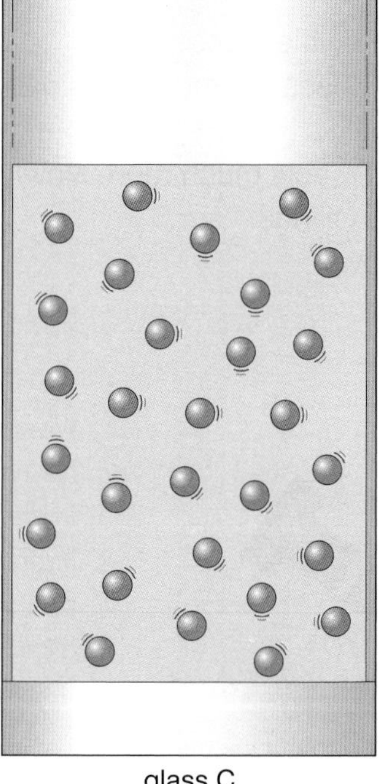

glass C

Many of the chemical reactions we see in school take place in solution. The substances are dissolved in water. For example, the acids you use are solutions. You might have noticed that their bottles are labelled with the acid's name. They also show a code, such as 0.5M, 1M or 2M. The code tells us how concentrated the acid is. The higher the number in front of M, the more concentrated the solution.

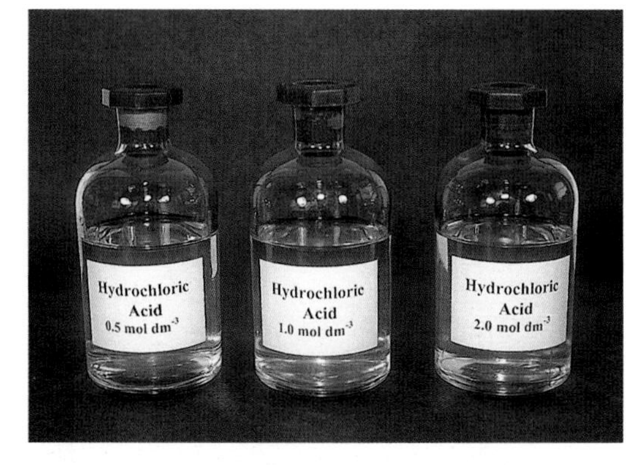

Bottles of dilute hydrochloric acid at different concentrations.

b) Which is the most concentrated acid in the photo?

c) What can you say about the number of acid particles in the 2M solution compared with the 1M solution?

Have you seen the reaction between hydrochloric acid and a marble chip (calcium carbonate)?
The calcium carbonate fizzes as carbon dioxide gas is given off.

Calcium carbonate reacting with dilute hydrochloric acid.

d) Given the 3 acids on the previous page, which will react fastest with a marble chip (calcium carbonate)?

As we increase the concentration, the rate of reaction increases.

Look at the diagram below:

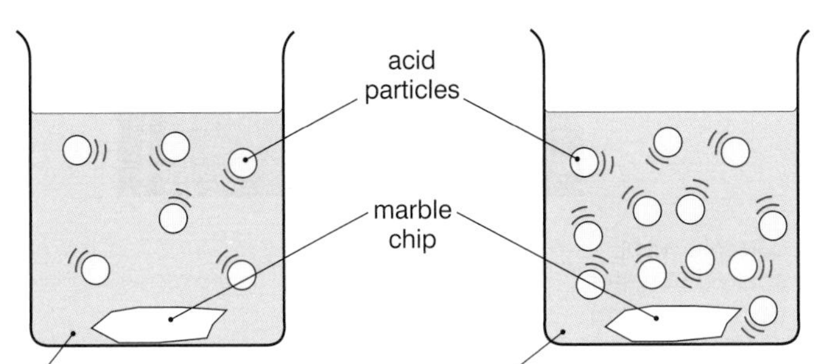

Remember that the particles in a liquid move around.
The acid particles bump into (collide with) the marble chip more often in the second beaker above. It's more 'crowded' in there.

e) Draw a beaker like the ones above for 0.5 M acid reacting.

Reactions with gases

We get the same effect if we increase the pressure in reactions between gases.
Look at the diagram opposite:
By squashing the gases, there are now the same number of gas particles but in a smaller volume. It gets more 'crowded'.
So there are more collisions in any given time.

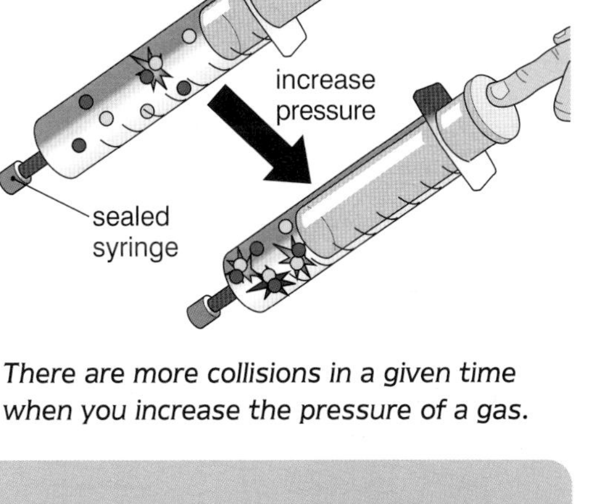

There are more collisions in a given time when you increase the pressure of a gas.

In reactions with gases, increasing the pressure increases the rate of reaction.

Remind yourself!

1 Copy and complete:

As we increase the ...... of a solution, its rate of reaction ...... This is because there are more p...... in the same ...... of solution. So we get more frequent ...... between them.

In reactions with ......, as we ...... the pressure, the rate of reaction increases.

2 The reaction between marble chips and hydrochloric acid is shown at the top of this page. The fizzing gets slower and slower, then eventually stops. There is still some marble chip left.

Explain these observations.

(Hint: Think about how the concentration of acid changes).

10e Effect of temperature

Can you imagine life without a fridge in your kitchen? What would happen to foods, such as milk, if we kept them at higher temperatures?
The reactions that make food go off are slowed down in a fridge.

a) Which other foods go off if we don't keep them cool?

Have you done any experiments in the lab where you heated up the substances reacting. For example, metals like tin or lead react with warm dilute acid.
Look at the reactions below:

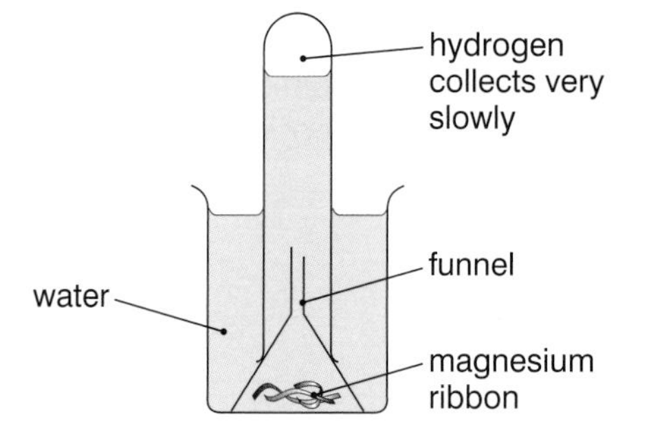

Magnesium ribbon reacting slowly with cold water.

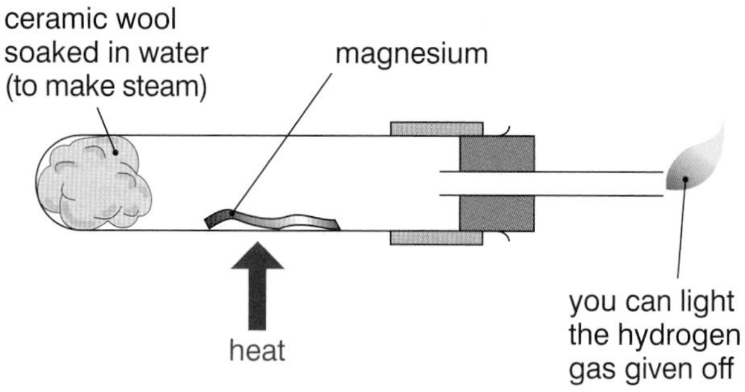

Magnesium ribbon reacting quickly with steam.

b) Does magnesium react faster with cold water or hot water?
c) If we raise the temperature, what happens to the rate of reaction?

Temperature has a big effect on the rate of a reaction. For many reactions, raising the temperature by just 10°C can double the rate of the reaction.

If we increase the temperature, we increase the rate of reaction.

So why the big increase?

Let's think back to the collision theory (page 126):
As we heat up the reacting particles, they gain more energy.
They start ***moving around faster***.
(It's a bit like you trying to walk on really hot sand.
You soon start hopping around!)

d) What will happen to the number of collisions in a set time if the particles are moving around faster?

e) What can you say about how hard they collide?

Look at the diagram below:

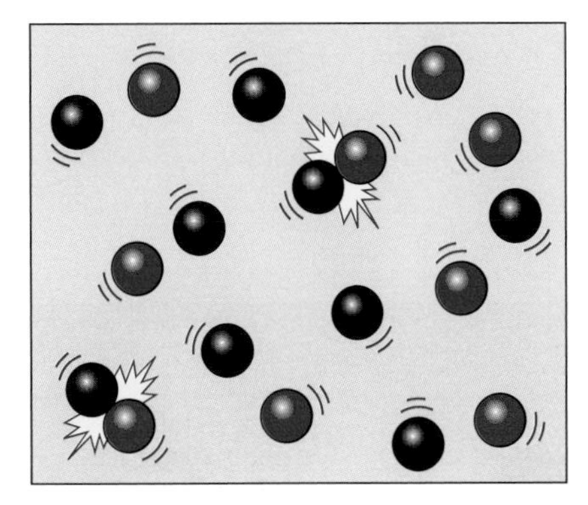

Reaction at 30°C.

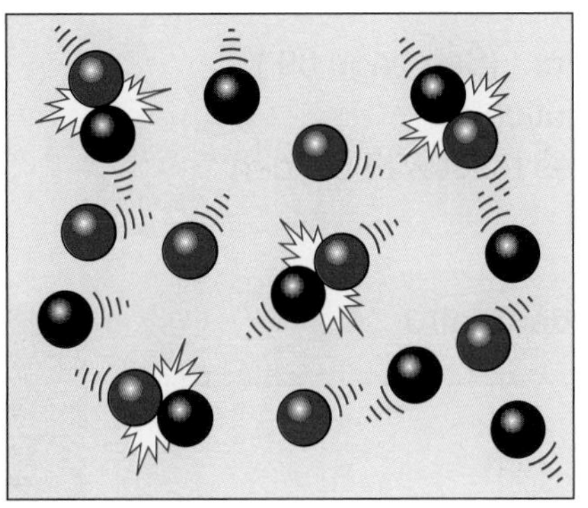

Reaction at 40°C.

A gentle collision produces no reaction.

On page 126 we said that not all collisions produce a reaction.
If the particles collide gently, they just bounce off each other.
But if they are moving around faster, more collisions will have enough energy to cause particles to react.
So there are two effects of raising the temperature.

As we raise the temperature the particles collide more often.

As we raise the temperature, the collisions are more effective. They are more likely to result in a reaction.

However, a harder collision may well produce a reaction!

Remind yourself!

1 Copy and complete:

As we increase the temperature, the rate of reaction ……

This is because the particles have more …… and move around …… This means that particles collide more …… and their collisions are more …… to result in a ……

2 Imagine that you are the owner of a fish and chip shop. You want to cook your chips as quickly as possible (because it's boring for customers waiting around!).

Write down two things you could do to speed up your chips.

10f Catalysts

Your 'mock' exams are sometimes the **catalyst** you need to make you start working harder!
In everyday life the word 'catalyst' means something that starts off a process.
In science:

> A **catalyst** is a substance that usually **speeds up** a reaction.
> At the end of the reaction, the catalyst is chemically **unchanged**.
> So they can be used over and over again.

You've probably heard of ***catalytic converters***. (See page 89.)
They are fitted to car exhausts to reduce pollution.
The catalysts help some of the pollutant gases react with oxygen before they are given out.

a) Which gases does a catalytic converter deal with?

The filler sets in a few minutes once the catalyst is mixed in.
Notice how little of the red catalyst you need.

We also get catalysts with some car fillers.
When a car has a small dent, or if you remove a little rust, you can repair it with a filler. The filler is a soft 'mushy' paste. It won't set until you add a little catalyst.
Then it reacts and turns hard in a few minutes.

Have you heard of the term 'peroxide blonde'?
A solution of hydrogen peroxide will bleach hair.
It breaks down, giving off oxygen gas.
Look at its reaction below:

⚠ hydrogen peroxide

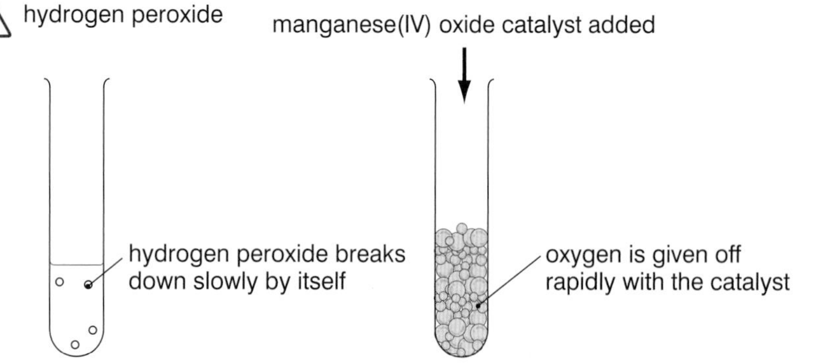

Manganese(IV) oxide is a catalyst for this particular reaction.
The equation with the catalyst can be shown as:

$$\text{hydrogen peroxide} \xrightarrow{\text{manganese(IV) oxide}} \text{water} + \text{oxygen}$$

> **Different reactions need different catalysts.**

b) Which gas is given off when hydrogen peroxide breaks down?

c) How could you get the manganese(IV) oxide back to use again? (Hint: It is insoluble in water.)

Catalysts let you make your products more quickly. So industry makes more money.

How do catalysts work?

You know from page 126 that colliding particles need a certain amount of energy before they can react. We call this minimum energy the **activation energy**.

A catalyst lowers the activation energy.

So with a catalyst, the particles need less energy to react. This means that more particles now have enough energy to start reacting. So reactions are faster.

You can think of it like a hurdle race.
If the hurdles are lower, more people can jump them and the race is over more quickly.

Remind yourself!

1 Copy and complete:

Most catalysts ...... up reactions but are not ...... themselves at the end of the reaction.

They work by ...... the energy needed for particles to ...... This minimum energy needed before a reaction can happen is called the ...... energy.

2 Do some research to find the catalysts used in each of these industrial processes:

a) Haber process to make ammonia.

b) Contact process to make sulphuric acid.

c) Making nitric acid.

d) Making margarine.

Summary

We can measure rates of reaction by looking at how quickly products are formed. We can also measure how quickly reactants are used up.

Rates of reaction are increased by:

- increasing the **surface area** (using small pieces) of solids,
- increasing the **concentration** of solutions,
- increasing the **pressure** of gases,
- increasing the **temperature,**
- using a **catalyst** (if you can find one for a particular reaction).

We explain rates of reaction using the **collision theory**.
Particles must collide, with enough energy, before a reaction can happen.
This minimum amount of energy is called the **activation energy**.

When we increase the concentration (or pressure in gas reactions), there are more particles in the same space so particles collide more often.

When we increase the temperature, the reacting particles gain more energy.
They move around faster, so collisions are more frequent.
The collisions are also more likely to produce a reaction because they collide with more energy.

Questions

1 Copy and complete:

We can increase the ...... of a chemical reaction by raising the ...... This makes particles ...... around more quickly so there are more ...... in a given time. When they collide, there is also more ...... of a reaction because collisions are harder.

When we increase the ...... of solutions, or the ...... of gases, we have more ...... in the same volume, so the rate of reaction ......

Catalysts lower the ...... energy of a reaction. More particles have ...... energy to react. The catalyst can be ...... over and over again.

2 Explain these statements:

a) Always keep milk in a fridge.

b) 3 cups of washing powder are needed instead of 1 cup if your washing is very dirty.

c) It takes a while to light coal on a fire but coal dust in a mine can explode.

d) Some glues come in two tubes; one big tube and a small tube labelled catalyst.

e) It takes longer to make toast when the grill is on a lower setting.

f) Industry invests a lot of time searching for catalysts to use when making new chemicals.

3 Imagine that everyone in your class was blind-folded. You walk around your classroom, bouncing off things you bump into. You represent particles of acid in solution.

The tables in your classroom represent marble chips. They are all packed together in the middle of the room.

Using this model:

a) Explain how the acid reacts with the marble chip.

b) How could you increase the surface area of the marble chip and explain what happens to the rate of reaction.

c) How could you increase the concentration of the acid? Why does this increase the rate of reaction?

d) What happens if you raise the temperature?

4 Look at the graph below:

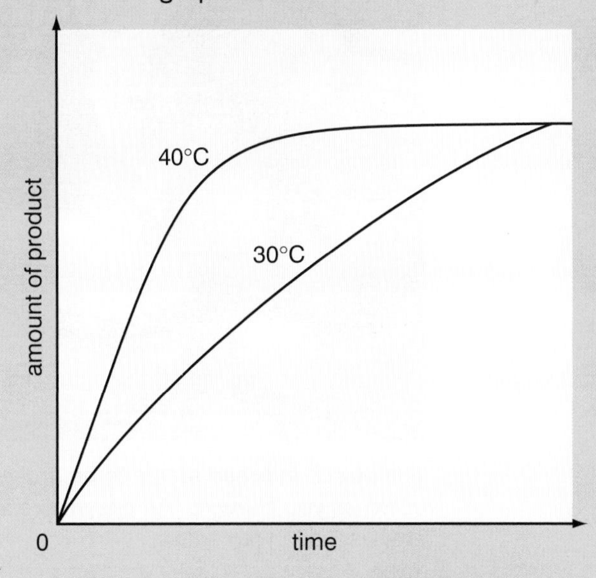

a) Was the reaction faster at 30°C or 40°C?

How can you tell from the graph?

b) Explain your answer to a) using the collision theory (talk about particles!).

c) When was each reaction going fastest?

d) Which reaction produced more product?

e) What would the line look like if you did the same reaction at 50°C?

5 A group of students looked at a reaction that produced a gas.
Here are their results:

Time (mins)	Volume of gas (cm^3)
0	0
1	20
2	29
3	37
4	38
5	44
6	45
7	45
8	45

a) Plot these results as points on a graph.

Put a circle around the point which seems to be a mistake. Then join the rest with a 'line of best fit' (in this case it's a curve).

b) How long did the reaction take to finish?

c) During which minute was the reaction fastest?

d) Draw the apparatus the students could have used to do their experiment.

e) The students repeated their experiment but used a catalyst. Sketch the line you might expect using the catalyst.

6 Look at the cartoon below:

Explain how the witch makes her potions so quickly!

CHAPTER 11 ENZYMES

11a What is an enzyme?

You've probably heard the word enzyme used in TV adverts for 'biological' washing powders. But what are these enzymes that can 'get rid of even the most stubborn stains!'.

a) Name 2 other uses of enzymes?

Enzymes are biological catalysts.
They are large protein molecules.

Enzymes are made in the cells of living things.
Without them we could not survive.
They allow all the chemical reactions that keep our bodies going to take place at body temperature.
For example, we use enzymes to help break down our food into small molecules. We also use them to build up big molecules.
In fact, as catalysts go, enzymes are brilliant!
Each enzyme molecule can help thousands of molecules to react every second.

The starch in bread starts to be broken down as you chew it. An enzyme called amylase is in your saliva.

b) Name an enzyme found in your saliva.

Here are some facts about enzymes:

Each enzyme only works for specific reactions.

Most enzymes work best at 'warm' temperatures.

Each enzyme works best at one particular pH.

Enzymes are very efficient catalysts.

Enzymes are molecules – they are not alive.

c) Why do enzymes work well at warm temperatures?
d) In which part of your body will you find enzymes that work best in acidic conditions?

Optimum (best) conditions

We can do experiments to show how well enzymes work in different conditions.

At different temperatures:

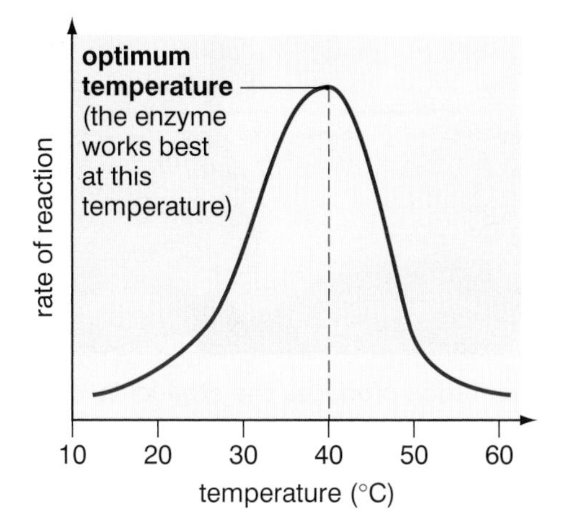

At different pH's:

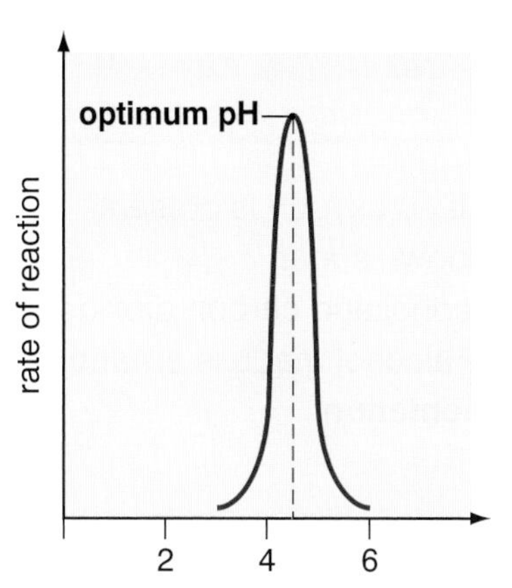

Use the graphs to answer these questions:

e) Enzymes have a temperature at which they work best. What do we call this temperature?

f) What pH value is best for the enzyme in the second graph?

Each enzyme has its own special shape.
Their shape matches the molecules they help to react.
Look at the diagram below:

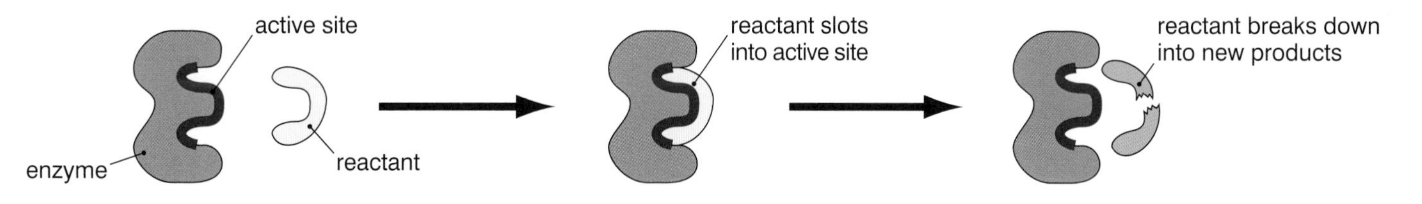

If you heat an enzyme above about 45°C, its shape changes.
It's a bit like shaking a delicate paper model. It loses its shape.
So enzymes don't work as well at higher temperatures.
We say that they are **denatured**.

Remind yourself!

1 Copy and complete:

Enzymes are biological …… found in cells. They work best at about ……°C and at one particular …… value.
If you heat them too much, their …… changes and they aren't so effective.

2 a) Sketch a graph to show how the effectiveness of an enzyme changes with temperature.

b) Imagine an enzyme as a paper model.

What happens to the enzyme as you raise its temperature?

11b Fermentation

Do you know anyone who brews their own beer or wine?
Home-brew kits always include a packet of **yeast**.

Fermentation produces the ethanol in alcoholic drinks.

a) Find out what brewers add yeast to when they make beer?

Yeast is a type of fungus. If no oxygen is present, enzymes in its cells break down sugar.
The sugar is turned into alcohol and carbon dioxide gas.
The chemical name for the alcohol made is **ethanol**.
The reaction is called **fermentation**.

Here is the equation:

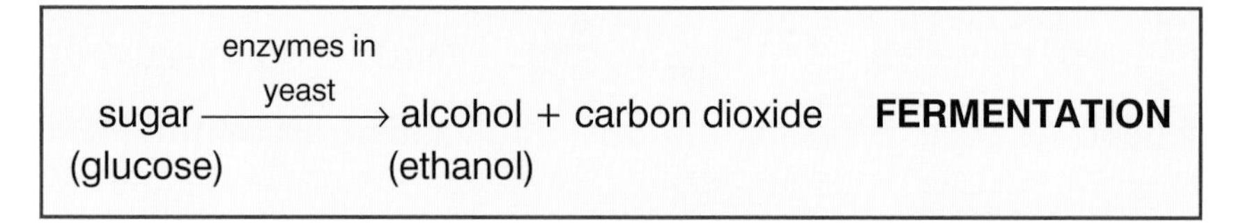

b) What is the reaction called in which yeast breaks down sugar to make alcohol?

c) Which gas is given off in this reaction?

d) What is the chemical name for the alcohol formed?

You can test the gas produced as shown below:

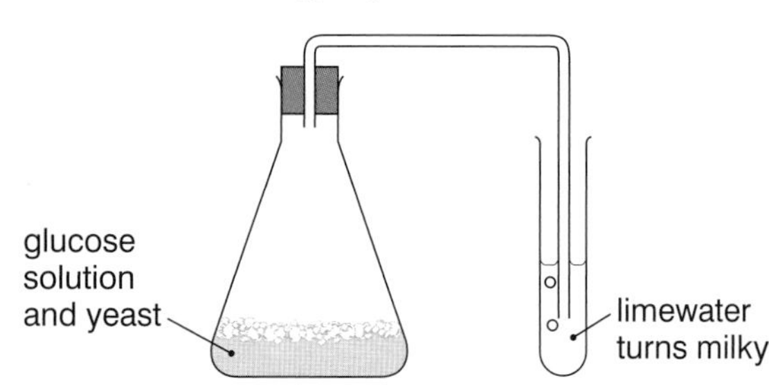

The flask is left near a radiator.

There was an unfortunate beast
Who fermented some sugar with yeast.
When he swallowed the brew,
His stomach, it grew
'Til BANG! CO_2 was released.

e) What is the test for carbon dioxide gas?

f) Why is the flask left near a radiator?

g) Why don't you boil the flask using a Bunsen burner?

The alcohol formed in the flask can't get much more concentrated than 15%. This is because the alcohol poisons the yeast. (Remember that yeast is a living thing – its enzymes aren't.) If we want stronger alcoholic drinks, such as vodka or gin, we have to distil the fermented mixture.

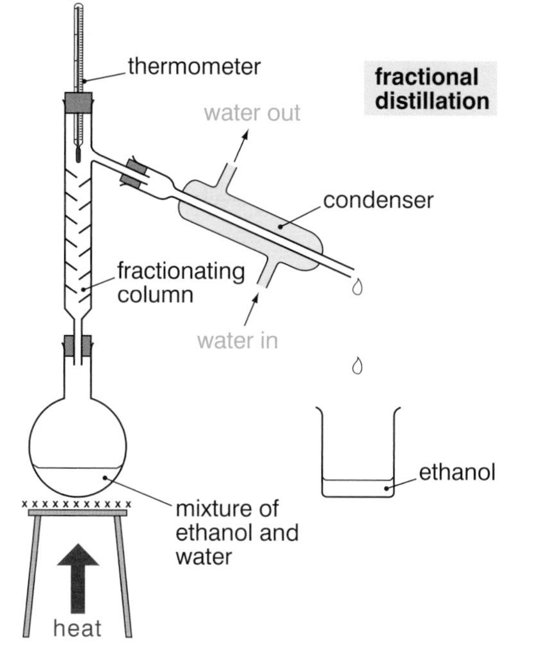

The alcohol (ethanol) boils at 78°C. Water boils at 100°C. So the alcohol boils off before the water and can be collected.

Fermentation is also important in the **bread making** industry. But in bread it's the carbon dioxide that's useful, not the alcohol. The carbon dioxide gas gets trapped in the dough and makes the bread 'rise'.
Look at the photo opposite:

Have you ever made bread yourself?
You have to leave the dough in a warm place for about an hour. This is called 'proving'.
Then it's ready to bake in the oven.

Fermentation helps bread dough to rise during its 'proving' stage.

h) Why is the dough left in a warm place? (Hint: Think about enzymes!)

Remind yourself!

1 Copy and complete:

...... in yeast cells can break down sugar into ...... (whose chemical name is ......) and ...... dioxide gas.

The reaction is called ...... It is used in the b...... industry to make alcoholic drinks. It also makes the ...... rise when making ......

2 a) Why do we have to distil a fermented mixture to make spirits?

b) Explain how the distillation works.

c) Draw a diagram of the apparatus you could use to distil a fermented mixture.

d) Draw a poster warning of the dangers of drinking too much alcohol.

▸▸▸ 11c More uses of enzymes

We have seen how the enzymes in yeast are used in the brewing and bread making industries.
These reactions have been used for thousands of years – long before people knew about enzymes!
They could also use the enzymes in bacteria to make cheese and yoghurt.

Do you like the slightly sour taste of natural yoghurt?
It doesn't taste as sweet as a glass of milk.
Bacteria turn the sugar in milk, called **lactose**, into an acid.
The acid is called **lactic acid**.

We have been using enzymes for thousands of years.

It is thought that yoghurt was first made in the Middle East.

a) Name 4 things made using enzymes that we eat or drink.
b) Name a sugar that we find in milk.
c) Why does yoghurt taste more sour than milk?

In industry people have realised that enzymes can save a lot of energy (and money).
Many industrial processes work at high temperatures and use transition metal catalysts.
Imagine if you could find an enzyme catalyst instead.
Just think of the savings you could make on energy.
(Remember that enzymes work best at about 40°C.
But many reactions in industry take place at about 400°C!)

d) How could enzymes save money in industry and help the environment?

So the search is on! For example, micro-organisms can be used to extract copper from slag heaps of waste from copper mines.
The waste would be too expensive to process normally.

Look at the table below:
It shows some other uses of enzymes.

Industry	Enzymes used to ……
medical	make drugs and treat cancer
biological detergent	break down stains (**proteases** attack proteins and **lipases** attack fats) they also soften fabrics (**cellulases** break down the 'bobbly' bits on clothes)
confectionery/sweets	break down starch syrup into sugar (glucose) syrup (**carbohydrases** are used) change glucose into fructose which is another sugar that is sweeter. So less sugar is needed to get the same sweetness – making it useful for slimming foods (**isomerase**) break down sucrose into glucose and fructose and make artificial sweeteners
meat	make the meat tender.
baby food	start off digestion of food (for example, proteases to break down protein)

e) Which enzyme can you find in some baby foods?

f) Which enzyme helps slimmers?

Enzymes in biological washing powders and liquids break down stains.

Baby foods contain enzymes to help the baby digest proteins by 'pre-digesting' them.

Remind yourself!

1 Copy and complete:

…… are used in many industries. For example,

- In …… making, bacteria turn lactose into …… acid
- Changing …… syrup into sugar syrup (using enzymes called ……)

2 Look at the table above:

a) Which enzymes break down protein?

b) Which enzymes break down fats?

c) Which enzyme changes glucose into fructose?

3 Do some research to find out how to make yoghurt or cheese.

Summary

- The chemical reactions in living cells are catalysed by **enzymes**. Enzymes are called biological catalysts.
- Enzymes are large protein molecules. Their complex shapes match the molecules they help to react.
- They work best at around 40°C (its optimum temperature). But over about 45°C, they are damaged and their special shape is changed. They are denatured and become much less effective.
- Each enzyme also works best at a particular pH value (called its optimum pH).

Remember that enzymes are not living things themselves – they are molecules.

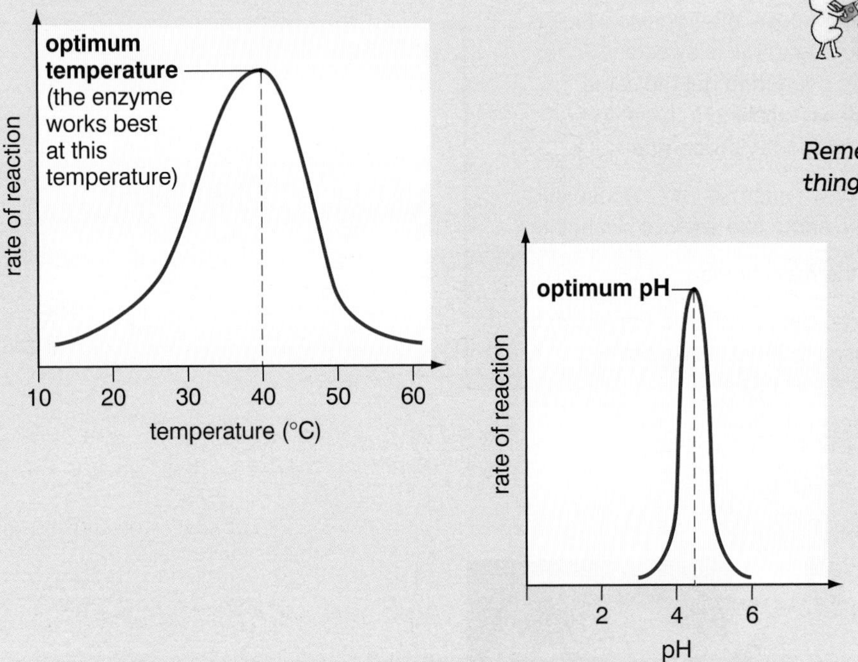

- **Fermentation** is the reaction in which yeast cells convert sugar (glucose) into alcohol (ethanol) and carbon dioxide gas.

  sugar (glucose) —(enzymes in yeast)→ alcohol (ethanol) + carbon dioxide

  The reaction is catalysed by enzymes in the yeast cells.
  We use fermentation to make:
  – the alcohol in wine and beer, and
  – the bubbles (of carbon dioxide) that make bread dough rise.

- Enzymes can save energy costs in industry because reactions can take place at relatively low temperatures.
  They are finding more and more uses. For example, enzymes are used:
  – in biological washing powders and liquids to break down stains,
  – in some baby foods to 'pre-digest' proteins,
  – to change starch syrup into sugar syrup,
  – to convert glucose into the less fattening sugar called fructose,
  – to make yoghurt (changing lactose in milk into lactic acid).

Questions

1 Copy and complete:

Enzymes are c…… found in the cells of living things. They are large molecules of …… They work best in warm rather than …… conditions. They also have an optimum …… value at which they are most effective.

Inside y…… cells, enzymes break down …… into …… and carbon dioxide. Brewers use this reaction, called ……, to make beer and w…… It is also useful when …… make …… as the …… …… gas makes the dough rise.

Other enzyme catalysed reactions include:

- making …… by changing the …… sugar in milk into lactic ……
- 'pre-digesting' p…… in some …… foods
- making the sugar called …… which is sweeter than glucose and is used in …… foods.

2 A friend's dad says, 'I can't wait till this home-brew wine is ready. It's going to be as strong as whisky because I put extra sugar in it!'

a) Explain why he will be disappointed.

b) Draw the apparatus you can use to separate alcohol from a fermenting mixture.

c) What is the process called in part b)?

When you visit your friend the following week, the impatient dad says, "I'm fed up waiting for this home-brew. I'm going to speed it up by heating it in a pan!"

d) Explain why this is not a good idea.

e) Why is it best to leave the home-brew to ferment in the airing cupboard?

3 A group of students looked at an enzyme reaction that gave off a gas.

They collected the gas produced in 2 minutes at different temperatures.

Here are their results:

Temperature (°C)	Volume of gas (cm^3)
20	8
30	21
40	32
50	20
60	5

a) Plot these results on a graph.

Draw a curve using a 'line of best fit'.

b) What does the phrase 'optimum temperature' mean?

c) What is the optimum temperature for the enzyme investigated here?

d) Name an enzyme reaction that gives off a gas.

e) Why do enzymes in your stomach work best at a low pH?

4 Look back through this chapter and find the names of all the enzymes mentioned.

Write down the reaction that each one speeds up. Put your answer in a table like this:

Name of enzyme	Reaction it catalyses

5 Why do industrial processes that involve enzymes save energy? What might cost a lot of money in these processes?

6 Design a cartoon that represents a molecule of protease doing its job in your body.

12a Energy and chemical reactions

Exothermic reactions

Did you know that you are using energy from a fuel as you are reading this book now? Your body gets the energy it needs to work from the food you eat. So you could say that food is your body's fuel.

A fuel is a substance that transfers energy stored in its molecules into useful energy for us to use.

a) Name 5 fuels that we use.

We release the energy from most fuels by burning them. We call these **combustion** reactions.
(Remember your work in Chapter 7 on the fire triangle and burning fossil fuels? See page 86.)

b) Think back to the fire triangle:
Which gas in the air reacts with a fuel as it burns?

The energy from the fuel in this dragster is transferred into heat, light and sound in its powerful engine (plus plenty of movement!).

We often use the energy transferred from fuels to heat things up. What fuel do you use to heat your home?

c) What happens to the temperature around a fuel when it burns?

We call reactions that give out energy **exothermic** reactions. The energy is often given out as heat energy, but you can also get light and sound. (Not forgetting the electrical energy we get transferred from the chemicals in cells and batteries.)

Exothermic reactions give out energy.

d) Name some reactions that give out heat, light ***and*** sound energy.

e) Try to list some uses of chemical reactions, other than combustion, that give out energy. (If you're stuck look ahead to page 147 for some ideas.)

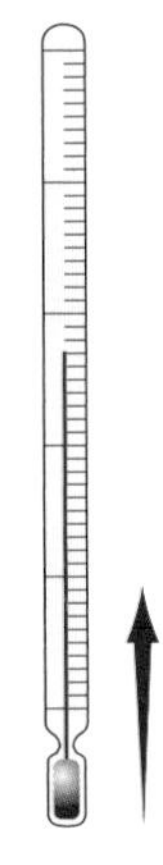

In exothermic reactions, the temperature goes up.

Endothermic reactions

The opposite of an exothermic reaction is an **endothermic** reaction. Therefore:

Endothermic reactions take in energy.

When you feel a test tube in which an endothermic reaction is taking place, it feels cold. The chemicals are taking in energy from your fingers (as well as the glass and the air). The temperature goes down.

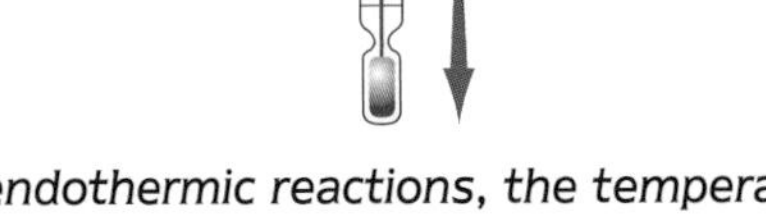

In endothermic reactions, the temperature goes down.

f) How does an endothermic reaction differ from an exothermic reaction?

Exothermic: *Heat 'exits' (is given out). The temperature outside goes up.*

Endothermic: *Heat 'enters' (is taken in). The temperature outside goes down.*

Remind yourself!

1 Copy and complete:

In an ...... reaction, energy is given out to the surroundings and the temperature ......

In an ...... reaction, energy is taken in from the surroundings and the temperature ......

2 Anna and Zara added two solutions in a beaker and recorded the temperature.

Here are their results:

Temperature before mixing = 21°C
Temperature after mixing = 33°C

a) What was the change in temperature?

b) Was the reaction exothermic or endothermic?

12b Useful exothermic reactions

We have already seen some of the useful exothermic reactions we get when fuels burn.
Look at some of the transport that relies on energy from exothermic reactions:

a) Do some research to find out the fuel used in each photo above.

All the fuels used above are hydrocarbons. (See page 78.)
When the fuels burn, we can show the reaction like this:

fuel + oxygen → carbon dioxide + water (+ HEAT) **An exothermic reaction**

But remember that heat is ***not*** a substance.
Therefore it is not a product in a chemical reaction.

Here is another useful exothermic reaction:

As soon as the car crashes, it sparks off
an exothermic reaction that produces nitrogen gas.
The gas fills the air bag in a split second.
The bag has small holes in it so that some of the gas escapes
when the body smashes into it. This makes it cushion the blow
even more.

b) Where do you find air bags fitted in modern cars?

You can also use the heat given out in a chemical reaction
to keep your hands warm in winter.
Have you ever seen these 'hand-warmers' in camping shops?

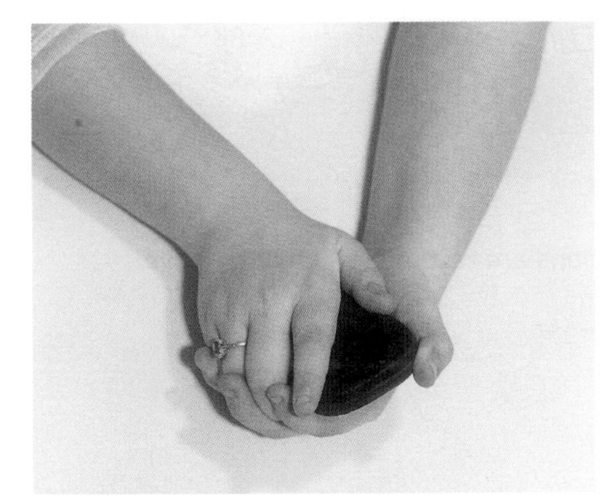

c) When might you find these hand-warmers useful?

Remind yourself!

1 Copy and complete:

A + B → C (+ HEAT)

shows an …… reaction.

2 Write an equation like the one in question 1 for an endothermic reaction.

3 a) Find out a use for an endothermic reaction.

b) Find out about the development of dynamite.

12c Reversible reactions

Many reactions are a bit like a one-way street.
The reactants (chemicals you start with) react together and form the products (the chemicals you end up with).
They only go one way. The products formed do not react to re-form the reactants.

reactants $\rightarrow$ products

a) Name a reaction that goes one-way only.

Other reactions are like a normal street. The traffic can go both ways.
So reactants form products. But the products can also react with each other and turn back into the reactants we started with.
These are called **reversible reactions**:

reactants $\rightleftharpoons$ products	***a reversible reaction***

Some reactions are 'one-way' only!

Other reactions are reversible – go both ways!

Imagine that the forward reaction is exothermic:

reactants $\rightarrow$ products (+ HEAT) EXOTHERMIC

then the backward reaction is endothermic:

products (+ HEAT) $\rightarrow$ reactants ENDOTHERMIC

The backward reaction will take in the ***same amount of energy*** as was given out in the forward exothermic reaction.

We can show this:

$$\text{reactants} \underset{\text{endothermic}}{\overset{\text{exothermic}}{\rightleftharpoons}} \text{products (+ HEAT)}$$

b) What would the general equation for a reversible reaction look like if the forward reaction was endothermic?

Under the right conditions, you can have both the reactants and products present together, and their amounts don't appear to change. In fact, they are reacting both ways – forwards and backwards – at the same rate. So it just looks like nothing is changing.

When this happens, we say that the reversible reaction is at **equilibrium**. Every time reactants turn into products, products turn into reactants to replace them.

c) Why does it look like nothing is happening when a reversible reaction is at equilibrium?

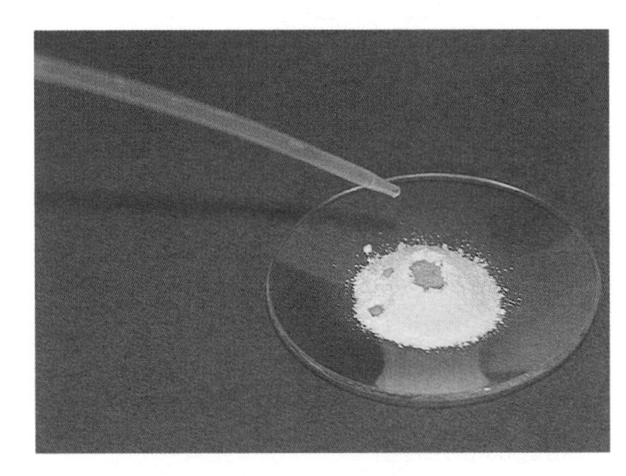

The test for water: white anhydrous copper sulphate turns blue.

Here are two examples of reversible reactions you might have seen:

Copper sulphate

Have you used copper sulphate in experiments before? Blue copper sulphate crystals are called **hydrated** copper sulphate. Its crystals have water molecules bonded inside them.

When you heat the blue crystals we get a white powder. The water is driven off. But if we add water to the white powder (called **anhydrous** copper sulphate), it turns blue again. We get hydrated copper sulphate back again. This is used as a **test for water**.

hydrated copper sulphate (+HEAT) $\rightleftharpoons$ anhydrous copper sulphate + water
BLUE crystals WHITE powder

Ammonium chloride

Ammonium chloride is a white solid. When you heat it, it turns into two colourless gases. The gases react with each other as they cool down to re-form the ammonium chloride.

ammonium chloride (+ HEAT) $\rightleftharpoons$ ammonia + hydrogen chloride
WHITE solid two COLOURLESS gases

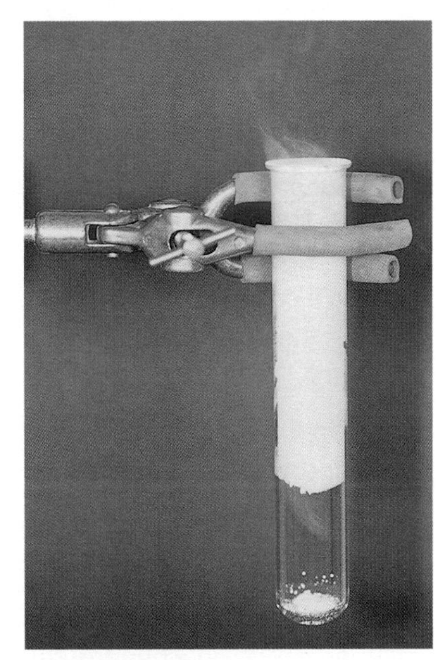

Ammonium chloride being heated.

Remind yourself!

1 Copy and complete:

Reactions that can 'go either way' are called ...... reactions.

If the forward reaction is exothermic, the ...... reaction will be ...... with the same amount of ...... involved each way.

2 a) What is the test for water?

b) Write an equation to show the reaction that takes place when testing for water.

c) Is the reaction in part b) exothermic or endothermic?

d) How can you test for ***pure*** water?

Summary

Reactions that give out energy, often as heat, are called **exothermic**.
The temperature of the surroundings rises.

Reactions that take in energy are called **endothermic**.
The temperature of the surroundings drops.

Some reactions are **reversible**.
The reactants form the products, but the products can also react together to re-form the reactants:

reactants ⇌ products

The test for water (white anhydrous copper sulphate turns blue) is an example of a reversible reaction.

If the forward reaction is exothermic, the backward reaction is endothermic.
If the forward reaction is endothermic, the backward reaction is exothermic.

In a reversible reaction:
The same amount of energy given out in one reaction will be taken in by the reverse reaction.

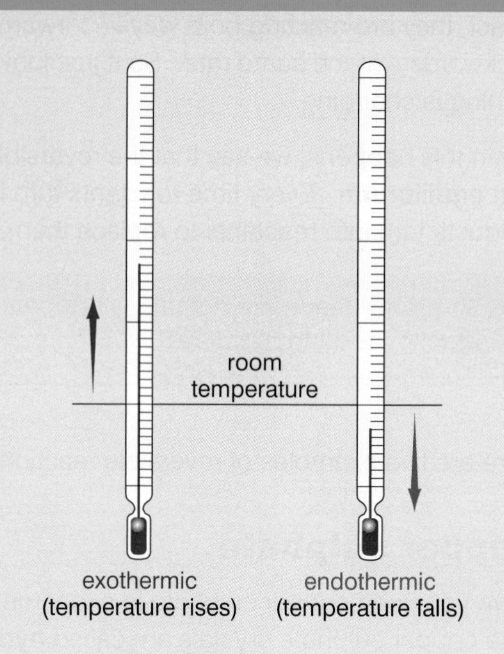

Questions

1 Copy and complete:

We call reactions that give out ...... exothermic. In these reactions the ...... of the surroundings goes ......

On the other hand, the ...... goes ...... in endothermic reactions.

In a ...... reaction, the reactants form the ...... and the ...... can re-form the reactants.

The same amount of energy will be ...... out as is ...... in when the reaction goes back and ......

2 A student heated hydrated copper sulphate in a test tube.

a) What does hydrated copper sulphate look like?

b) What is given off as the test tube is heated?

c) What do we call the solid left behind?

The student was given two colourless liquids, but only one contained water.

d) How could she use her experiment above to find out which had water present?

3 Wes and Cara did an experiment adding different powders to the same volume of dilute acid. They took the temperature of the acid before they added the powder. Then they recorded its highest temperature reached during the reaction.

Here are their results:

With **powder A**:
Temperature before = 19°C
Maximum temp. = 22°C

With **powder B**:
Temperature before = 19°C
Maximum temp. = 25°C

With **powder C**:
Temperature before = 20°C
Maximum temp. = 22°C

With **powder D**:
Temperature before = 20°C
Maximum temp. = 18°C

a) Which powders reacted in exothermic reactions and which were endothermic?

b) Which reaction gave out most heat energy?

c) Explain why an endothermic reaction feels cold if you touch the beaker used in the experiment.

4

Coal gets very hot when it burns.

a) Is this an exothermic or endothermic reaction?

b) Describe the coal's energy changes from before being lit to when it's blazing away.

5 A group of students took the temperature of a reaction between two solutions as it went along.

Here are their results:

Time (mins)	Temperature (°C)
0.0	20
0.5	24
1.0	28
1.5	32
2.0	36
2.5	38
3.0	36
3.5	34
4.0	32

a) Plot these results on a line graph. (Put time along the bottom.)

b) What was the temperature of the room?

c) What was the maximum change in temperature recorded?

d) How long did the reaction take to reach its maximum temperature?

e) Was the reaction exothermic or endothermic?

f) How could the students have used a computer to monitor this experiment?

What are the advantages of using computers to monitor some experiments?

6 When you heat ammonium chloride in a test tube it decomposes into two colourless gases.

a) Name the gases.

The ammonium chloride forms again on the cooler sides higher up the test tube.

b) Write a word equation to explain the changes in the test tube.

Concentrated ammonia solution gives off ammonia gas. It can be used to test for hydrogen chloride gas.

c) Describe what you would see in the test.

7 a) What do we mean by the term 'combustion reaction'?

b) Make a list of 5 fuels.

c) Give a use of each fuel you named in part b).

d) Photosynthesis is the process in which plants make glucose (a sugar). It is an endothermic reaction.

Where do plants get the energy needed for the reaction to happen?

CHAPTER 13 Fertilisers

13a Why we need fertilisers

Do you enjoy a bowl of cereal for your breakfast?
Our breakfast cereals are often based on wheat products.

The crops used in this breakfast cereal have been grown with the help of fertilisers.

a) Check some cereal packets and see which other crops are used in breakfast cereals.

Have you seen large fields of wheat being harvested at the end of summer? When wheat and other crops are harvested, they take with them minerals that contain essential elements from the soil.

As plants grow, they absorb the minerals that they need. If a plant goes on to die and decompose in the field it grew in, then the elements are replaced in the soil.

But on a farm, this natural cycle is broken. The essential elements don't get a chance to get back into the soil. And that's why farmers need to add **fertilisers** to their fields.

These crops do not get a chance to replace nitrogen back into the soil.

b) How do plants that die naturally replace essential elements in the soil?

One of the main essential elements that plants need to grow is **nitrogen**. But you might think that there should be no need to replace that element; isn't most of the air made up of nitrogen? However, nitrogen gas in the air can't be used by most plants. They need the nitrogen to be in a soluble compound of nitrogen. Then their roots can draw it up from the soil dissolved in water.

c) What percentage of the air is nitrogen gas? (See page 96.)

d) Find out 3 plants that can use nitrogen directly from the air. (These have nodules on their roots that absorb nitrogen.)

e) How do other plants get their essential nitrogen?

The chemical industry makes the fertilisers that help to feed the world. (See page 157.)

The trouble with fertilisers ...

Fertilisers must dissolve in water for plants to use them. Only then can the plants absorb the essential elements in minerals through their roots. The soluble nitrogen compounds used are often **nitrates**.

But the solubility of fertilisers creates problems for us. When it rains, the fertilisers can be washed (leached) out of the soil. They drain away into groundwater and find their way into streams and rivers.

We now know that this is affecting our drinking water in some places.
People are worried about this causing stomach cancer.
Some also blame nitrates in water for 'blue baby' disease.
This is when a new-born baby's blood is starved of oxygen.

Water companies are trying to remove the nitrates. However, some people are still getting water with more nitrates in than they should.

f) What do you think about paying more for your water to make sure that most nitrates are removed from your drinking water?

The fertilisers washed into the rivers also cause a problem called **eutrophication**. The river plants and algae grow quickly. When the algae die, bacteria thrive as they decompose them. They use up the oxygen dissolved in the water.
This means that fish and other water animals cannot get the oxygen they need and soon die.

g) What do you think happens to plants that grow on river beds when algae completely cover the surface?

This river has been affected by fertilisers washed into it from fields.

Remind yourself!

1 Copy and complete:
Although almost ...% of the air is made up of nitrogen, most ...... cannot use this directly.
They need to take it in as s...... nitrogen compounds through their ......
These compounds can p...... our water as they get l...... out of the soil.

2 a) What are the most common nitrogen-based fertilisers called?

b) Why are people worried about these compounds getting into:
i) our drinking water
ii) rivers?

3 Write a magazine article that is called 'Fertilisers: Heroes or villains?'.

13b The Haber process

As you know, most plants cannot use nitrogen directly from the air.
But chemists have found a way to turn nitrogen from the air into a soluble compound called **ammonia** (NH_3). This is a gas.
Then we use the ammonia to make solid fertilisers that are easy to spread on fields.

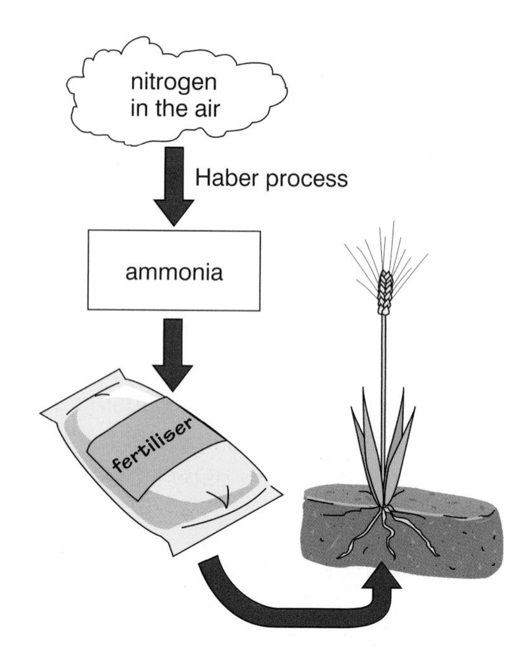

Nitrogen from the air is converted into useful nitrogen for crops.

a) Nitrogen gas itself is not soluble in water. Name a soluble gas that contains nitrogen.
b) What is the chemical formula of this gas?

The process to change nitrogen into ammonia was discovered in 1909. A German chemist called Fritz Haber was the first person to succeed, although he only managed to make 100 g.

nitrogen + hydrogen $\rightleftharpoons$ ammonia

c) Which elements react together to make ammonia?
d) What does the $\rightleftharpoons$ sign mean? (See page 148.)

Fritz Haber (1868–1934) got the Nobel Prize for Chemistry in 1918 for making ammonia from nitrogen and hydrogen. He was forced to leave Germany in 1933 when Hitler came to power. He died soon afterwards.

It was vital for Germany to make lots of ammonia.
At that time they were preparing for the First World War.
They knew that they would need nitrogen compounds to make explosives, and fertilisers to feed their people.
They were importing nitrogen compounds from South America but realised these could be cut off once war started.

So they raced to 'scale up' the process to make ammonia.
An engineer called Carl Bosch finally built the steel vessels needed for the high pressures necessary (not before his first effort had blown up!).

The scientists also tried more than 6500 experiments to find the best catalyst for the process.
Iron is the main catalyst used.

Carl Bosch (1874–1940) was awarded the Nobel Prize for Chemistry in 1931 for his work on high-pressure reactions.

e) Why was it important for Germany to find a way to make their own nitrogen compounds?
f) What different roles did Haber and Bosch play in discovering the process?

Look at the diagram below which shows the **Haber process**:

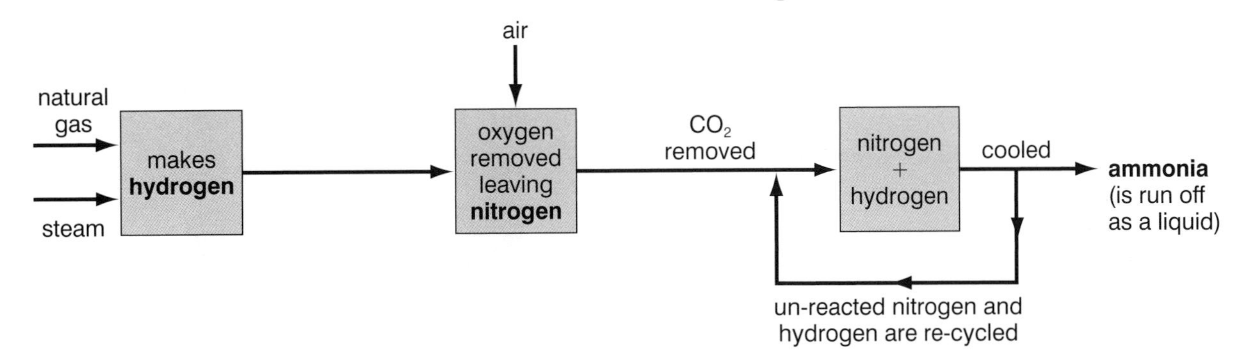

The materials we need to get the reactants are called the **raw materials**. The raw material for nitrogen is air. We get hydrogen from natural gas.

The reaction is reversible. So as nitrogen and hydrogen react together to make ammonia, the ammonia breaks up and re-forms nitrogen and hydrogen.
Therefore the choice of conditions is very important.
We want to make as much ammonia as quickly as possible.
Here are the conditions chosen:

The conditions used for the **Haber process** are:
- **a temperature of about 450°C,**
- **a pressure of about 200 atmospheres,**
- **a catalyst of iron.**

Notice from the flow diagram above that any un-reacted nitrogen and hydrogen is recycled.
The ammonia gas made is cooled down into a liquid and collected.

g) Why is the ammonia cooled down?

Remind yourself!

1 Copy and complete:

...... is made in the Haber process.

The reaction is r......:

nitrogen + ... ⇌ ammonia

The conditions chosen are about ......°C, ... atmospheres and a catalyst of ......

2 a) Where do we get the nitrogen gas needed for the Haber process?

b) Where do we get the hydrogen for the process from?

c) How is ammonia removed from the mixture of gases that pass out from the reaction vessel?

13c Making fertilisers

Most of the ammonia from the Haber process is used to make fertilisers.
Ammonia is an alkaline gas so isn't much use as a fertiliser itself.
But because it is alkaline, it ***reacts with acids***. (See page 55.)
We can then use the ***salt made*** as a solid fertiliser.

This is ammonium nitrate – a nitrogen-based fertiliser. It is made by reacting ammonia with nitric acid in a neutralisation reaction.

a) Why isn't ammonia used as a fertiliser very often?

Making nitric acid

Ammonia is actually used as a starting material for making **nitric acid (HNO_3)**. We use this to react with more ammonia to make a fertiliser. (See next page.)

Look at the diagram showing the process used to make nitric acid:

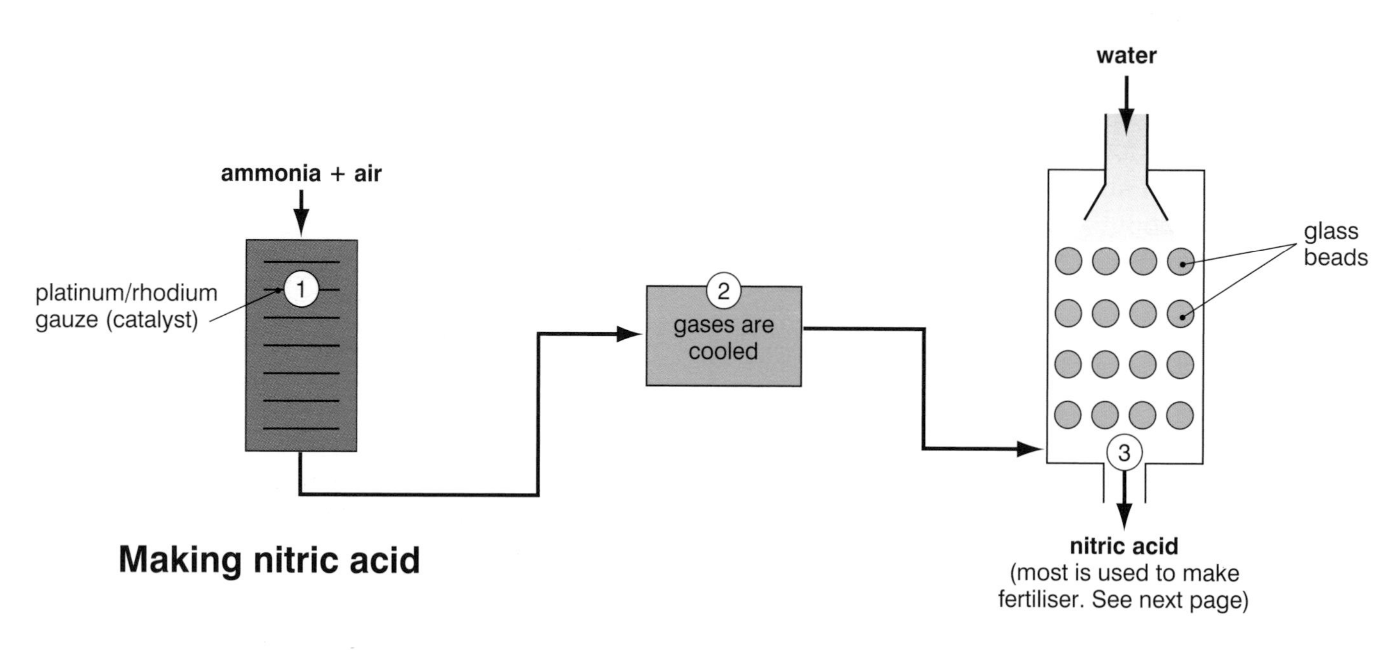

Making nitric acid

Step 1

The ammonia reacts with oxygen in the air.
We say that the ammonia is **oxidised**.
This reaction won't happen without a catalyst.
Layers of gauze made from precious metals (platinum mixed with some rhodium) are used as the catalyst.
The temperature is 900°C in the reaction vessel.

$$\text{ammonia} + \text{oxygen} \xrightarrow{\text{platinum / rhodium}} \text{nitrogen monoxide} + \text{water}$$

Step 2

The nitrogen monoxide is then mixed with air and cooled. It forms nitrogen dioxide:

nitrogen monoxide + oxygen → nitrogen dioxide

Step 3

Finally, the nitrogen dioxide and more oxygen react with water to make nitric acid:

nitrogen dioxide + oxygen + water → nitric acid

b) What is used as a catalyst in the process to make nitric acid?

c) Why is it worthwhile using precious metals in the process?

Making ammonium nitrate

We now have the starting materials to make a fertiliser. We have ammonia solution (an alkali) and nitric acid. These react together to make the salt called ammonium nitrate.

ammonia solution + nitric acid → ammonium nitrate + water

This is a **neutralisation** reaction. (See page 55.)

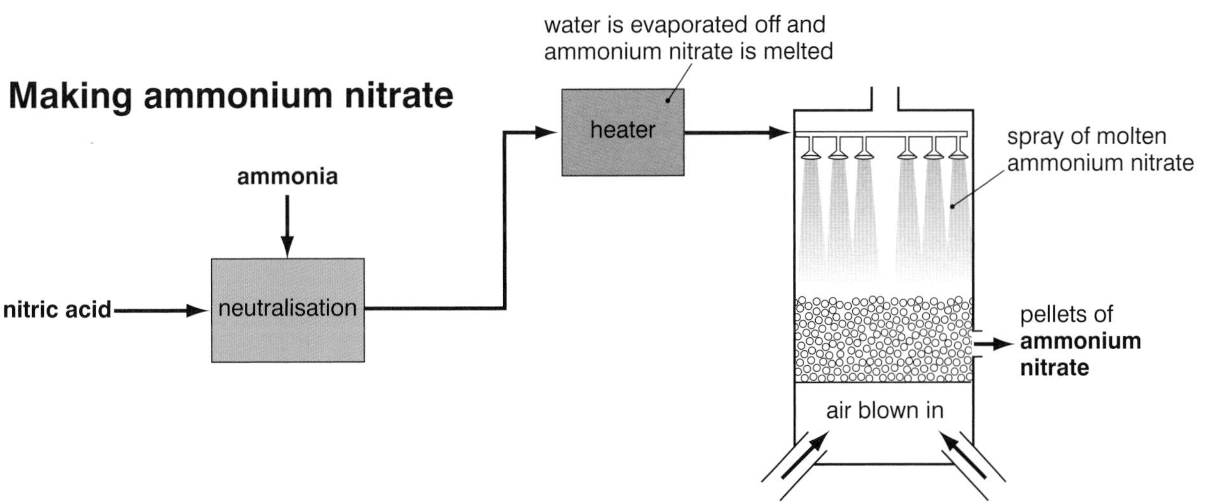

Remind yourself!

1 Copy and complete:

...... acid is made by oxidising ammonia on a ...... / rhodium catalyst. The ...... monoxide formed is ...... down and reacted with more ...... and water to make the acid.
If the acid is n...... by ammonia we get the fertiliser called ....

2 a) Write word equations to show the 3 steps in the manufacture of nitric acid.

b) Write a word equation to show the reaction in which nitric acid is converted to a fertiliser.

c) Why is ammonium nitrate more convenient to use as a fertiliser than ammonia?

13d How much nitrogen?

Imagine that you are a farmer buying some fertiliser.
You want to add nitrogen to your soil to help your crops grow.
Naturally you would want good value for money.
We can work out the percentage of nitrogen in the fertiliser to see which provides most goodness for the crops.

You might be given a choice of 3 fertilisers:

Fertiliser	Its chemical formula
ammonium nitrate	NH_4NO_3
ammonia	NH_3
urea	$CO(NH_2)_2$

Each element has a R.A.M.!

We know the mass of each atom compared to the lightest of all atoms, hydrogen.
We can make a scale in which hydrogen (H) is given the value 1. On this scale, nitrogen weighs 14.
It is 14 times as heavy as hydrogen.

These numbers are called the **relative atomic masses**.
They are given the symbol A_r.
Look at some values opposite:

Atom	Relative atomic mass (A_r)
hydrogen	1
carbon	12
nitrogen	14
oxygen	16

a) How many times heavier than a hydrogen atom is a carbon atom?
b) Which element in the table has the heaviest atoms?

Given these values of relative atomic mass, let's look at the fertilisers we can choose from the table above.
We can now work out the **relative formula mass (M_r)** of each fertiliser. This tells us the mass of compounds.
We simply ***add up the masses of all the atoms*** in a compound.
Let's take ammonia (NH_3) as an example:

We know an ammonia (NH_3) molecule is made up from 1 nitrogen atom and 3 hydrogen atoms.
So the relative formula mass is:

$$\begin{aligned} 1\text{ N} &= (1 \times 14) = 14 \\ +\ 3\text{ H} &= (3 \times 1) = \underline{3} \\ & \qquad\qquad \underline{17} \end{aligned}$$

The relative formula mass (M_r) of ammonia is 17.

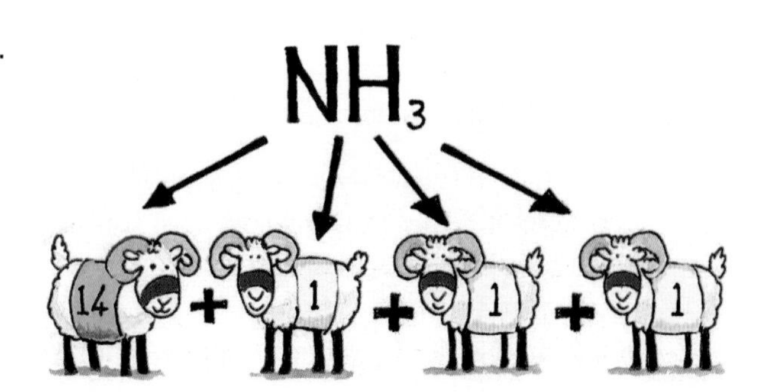

Add up the R.A.M.s to get the Relative Formula Mass.

For urea, whose formula is $CO(NH_2)_2$, we will have to add up:

1 carbon, 1 oxygen, 2 nitrogens and 4 hydrogens
$12 + 16 + (2 \times 14) + (4 \times 1)$

So the relative formula mass (M_r) of urea is $12 + 16 + 28 + 4 = 60$

If a formula has brackets, multiply all the atoms inside the brackets by the number outside. So $(NH_2)_2$ shows 2 N's and $(2 \times 2) = 4$ H's.

c) Work out the relative formula mass of ammonium nitrate, NH_4NO_3.
Show your working out.
(Did you get 80?)

Finding the percentage of nitrogen in a fertiliser

Once we know the relative formula mass of each fertiliser,
we can find the percentage of nitrogen in each one.
Let's look at ammonia again.
Its mass is 17. Of that, 14 is nitrogen.

So the percentage of nitrogen in ammonia (NH_3) is:

$$\frac{\text{Mass of nitrogen}}{M_r \text{ of ammonia}} \times 100 = \frac{14}{17} \times 100 = 82.4\%$$

And the percentage of nitrogen in urea –$CO(NH_2)_2$ – is:

$$\frac{\text{Mass of nitrogen}}{M_r \text{ of urea}} \times 100 = \frac{28}{60} \times 100 = 46.7\%$$ (Notice that there are 2 N's in urea.)

d) Work out the percentage of nitrogen in ammonium nitrate, NH_4NO_3.
Show your working out.
(Did you get 35%?)

e) Which of the 3 fertilisers has the highest percentage of nitrogen?

f) What other things would you have to think about before deciding which fertiliser to buy?

So if you want to find the percentage of any element in a compound, you can use this formula:

$$\frac{\text{Mass of element}}{\text{Relative formula mass of compound}} \times 100 = \textbf{\% of element present}$$

Remind yourself!

1 Copy and complete:

Atoms of different elements have different ......
We compare these using ...... atomic ......
(symbol ......). We add these up to get the relative ...... mass of a compound (symbol ......).

2 a) What is the relative formula mass of:
i) CO ii) CO_2 iii) N_2H_4 iv) CH_3NH_2
(Use A_r values from the previous page.)

b) Work out the percentage of oxygen in
i) CO and ii) CO_2.

c) What percentage of nitrogen is in N_2H_4?

Summary

Although almost 80% of the air is nitrogen gas, most plants can't use this directly to help them grow. So we add nitrogen-based **fertilisers** to the soil. These are soluble compounds that can be absorbed through the roots of a plant.

However, these fertilisers do cause pollution in our water supplies and in rivers.
In rivers and lakes, fertilisers cause algae to thrive.
When they die, the micro-organisms that decompose the algae use up the oxygen dissolved in the water.
Then fish and other water animals die.
This is called **eutrophication**.

Nitrogen is converted to ammonia in the **Haber process**.

nitrogen + hydrogen $\rightleftharpoons$ ammonia

The catalyst used is iron. The temperature is about 450°C and the pressure about 300 atmospheres.

Some of the ammonia is used to make **nitric acid**. Ammonia is oxidised to nitrogen monoxide. This is then cooled down and reacted with more oxygen and water to make the nitric acid.

We can then use the neutralisation between nitric acid and more ammonia to make the fertiliser called **ammonium nitrate**.

Questions

1 Copy and complete:

......% of the air is nitrogen but most ...... can't absorb this through their ......

In the ...... process, nitrogen is reacted with ...... to make ammonia an a r...... reaction. The process operates at about ......°C and at a pressure of about ...... atmospheres. The catalyst used is ......

Ammonia can then be used to make ...... acid.

It is first ...... to make ...... monoxide. This is ...... down and reacted with more ...... and water to make ...... acid.

Ammonia and nitric acid react together to make the fertiliser ammonium ...... in a ...... reaction.

2 Look at this limerick:

There was a young man from Leeds
Who swallowed a packet of seeds.
With N, P and K
And pesticide spray,
He's now fully covered in weeds.

a) What are the essential elements N, P and K?

b) Which essential elements do the following compounds give to plants?

i) ammonium nitrate – NH_4NO_3

ii) potassium chloride – KCl

iii) ammonium phosphate – $(NH_4)_3PO_4$

iv) ammonium sulphate – $(NH_4)_2SO_4$

3 a) Explain why crops need fertilisers to help them grow.

b) Name 2 nitrogen-based fertilisers.

c) How do fertilisers find their way into groundwater, streams and rivers?

d) Why are people worried about drinking water with too much nitrate dissolved in it?

e) Explain what happens to rivers when fertilisers get into them.

f) What are your views on genetically modified crops?

4 This question is about the Haber process.

Look at the graph below:

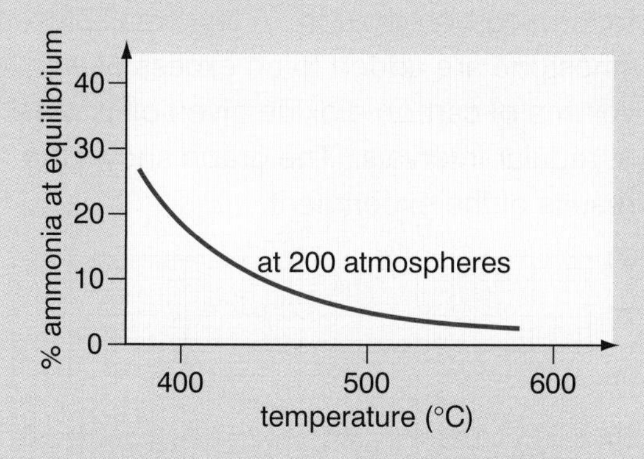

a) What happens to the percentage of ammonia in the reacting mixture as the temperature rises?

Look at the graph below:

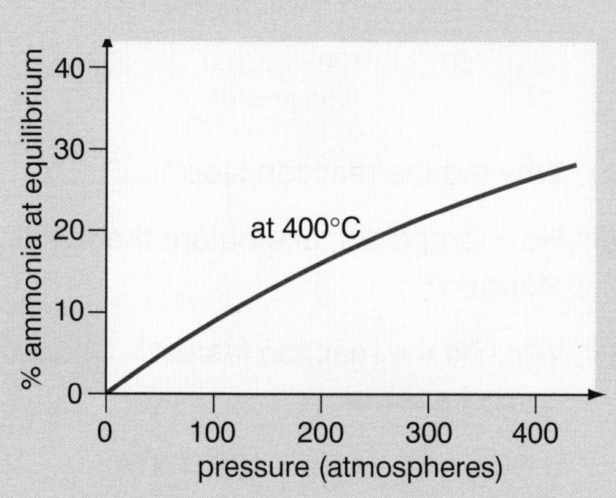

b) What happens to the percentage of ammonia in the reacting mixture as the pressure rises?

5 This question is about making nitric acid.

Copy and complete this flow diagram of the process:

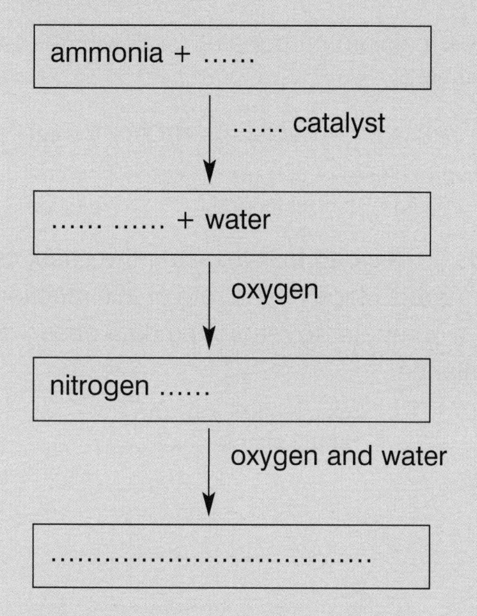

6 Work out the relative formula mass of these compounds:

Show your working out.

a) H_2O

b) HCl

c) H_2S

d) $CaCO_3$

e) HNO_3

f) $(NH_4)_2SO_4$

g) C_4H_{10}

h) $C_6H_{12}O_6$

(The relative atomic masses you need are:

H = 1, O = 16, Cl = 35.5, S = 32, Ca = 40, C = 12, N = 14.)

7 Work out the percentage of carbon in the following compounds:

Show your working out.

a) CH_4

b) HCN

c) CO_2

d) Na_2CO_3

Further questions on Chemical reactions and their useful products

Rates of reaction

1 Marble chips (calcium carbonate) react with dilute hydrochloric acid.

$CaCO_3(s) + 2HCl(aq) \rightarrow CaCl_2(aq) + H_2O(l) + CO_2(g)$

calcium carbonate + hydrochloric acid → calcium chloride + water + carbon dioxide

Some students decided to investigate the effect of the size of marble chips on the rate of this reaction. They used a computer to record the data observed from the balance.

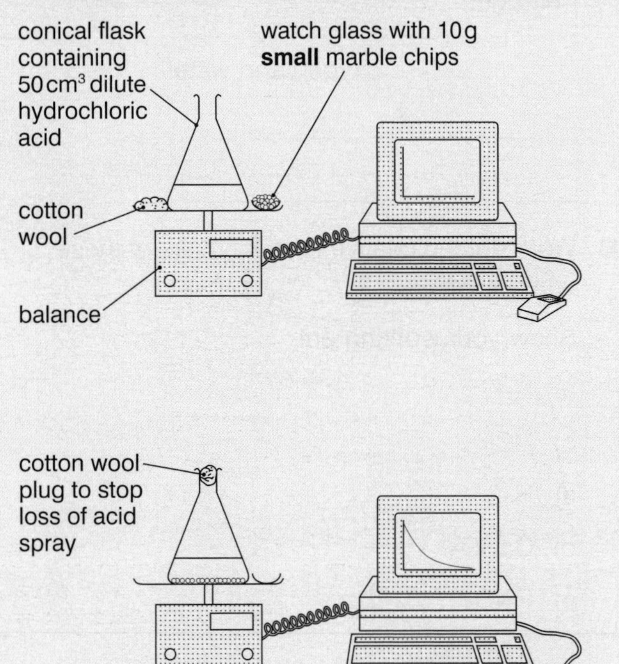

The students repeated the experiment three times using different sizes of marble chips **A**, **B** and **C**. All other conditions were kept the same. The graph is a computer printout of all three sets of results.

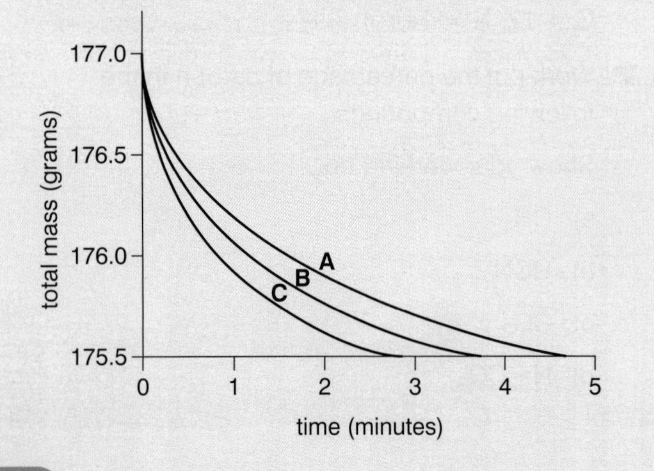

(a) Explain why there was a loss in mass as each reaction took place. (1)

(b) (i) How does the rate of reaction change with time? (1)

(ii) Explain why this change in the rate of reaction took place. (1)

(c) Which of the curves, **A**, **B** or **C**, shows the results for the smallest marble chips? (1)

(d) Explain, in terms of particles, why the size of the marble chips changes the rate of reaction. (2)

(AQA 2000)

2 **(a)** Limestone reacts with dilute hydrochloric acid to form carbon dioxide. A few small pieces of limestone are added to an excess of acid. The volume of carbon dioxide given off is measured at regular intervals. The graph shows the results of the experiment.

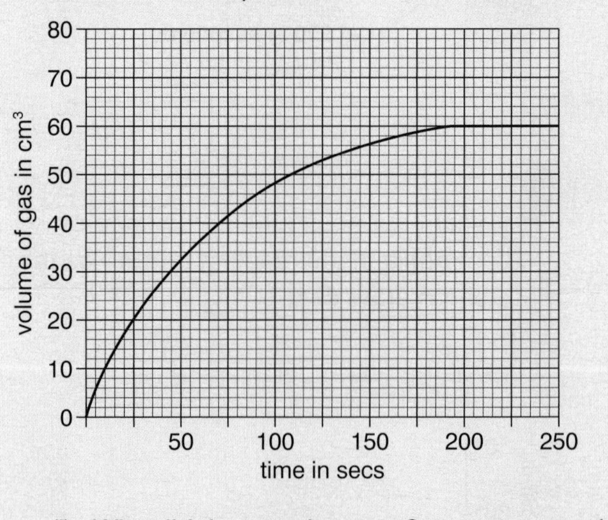

(i) Why did the reaction stop? (1)

(ii) How long did it take before the reaction stopped? (1)

(iii) When is the reaction fastest? Choose the correct answer.

0–50 s 50–100 s 100–150 s

150–200 s 200–250 s (1)

(b) Describe THREE ways in which a reaction can be made to go faster. (3)

(EDEXCEL 1999)

3 Some students investigate the rate of a reaction.

They use 40 cm^3 of a sodium thiosulphate solution in a beaker.
They stand the beaker over a black **X** on a piece of card.
Then they add 5 cm^3 of dilute hydrochloric acid and start the clock.

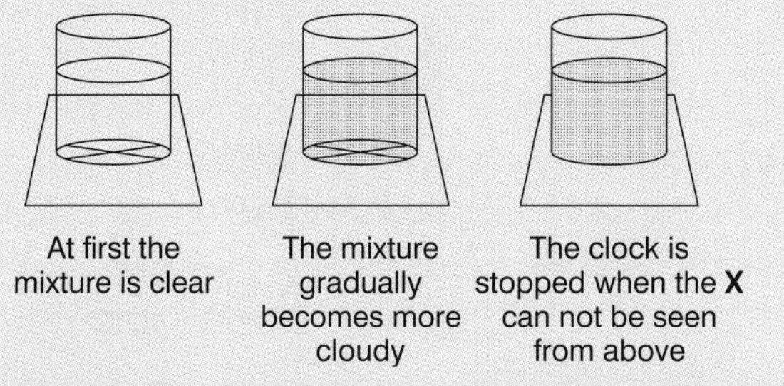

The students repeat the experiment twice using different strengths (concentrations) of sodium thiosulphate solution as shown in the table below.
All three experiments are carried out at 20°C.

Experiment	Strength of thiosulphate solution	Time for X to disappear (s)
1	Full strength	63
2	Half strength	124
3	Quarter strength	255

(a) What do these results tell us about the rate of this reaction? (2)

(b) If the reaction had been carried out at 40°C, it would have taken place more quickly.
Explain why, in terms of particles. (3)

(AQA 2001)

4 Some types of filler go hard after a catalyst is added from a tube. A manufacturer tested this reaction to see what effect the amount of catalyst had on the time for the filler to harden. The results are shown in the table.

VOLUME OF CATALYST ADDED TO FILLER (cm^3)	TIME FOR THE FILLER TO HARDEN (minutes)
1	30
2	15
3	10
4	7
6	4

(a) Draw a graph of these results.

(b) Use your graph to suggest the time taken for the filler to harden using 5 cm^3 of catalyst. (1)

(c) What is the effect of the catalyst on the rate of this reaction? (1)

(AQA 1999)

5 Ann and Nick investigate the reaction between sodium thiosulphate and hydrochloric acid.
A yellow solid is made during the reaction.
Look at the diagram.

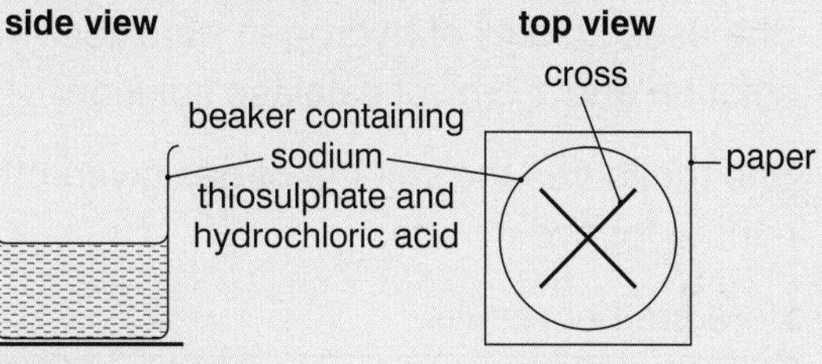

Ann and Nick look down at the cross. As the yellow solid is made, the liquid in the beaker gets cloudy.
After a time they cannot see the cross.
Ann and Nick measure this time.
They do four experiments.
They use four different concentrations of sodium thiosulphate solution (**A**, **B**, **C** and **D**).
They do all the experiments at 20°C.
The table shows their results.

Concentration	Time taken for cross to disappear in seconds
A	42
B	71
C	124
D	63

(a) Look at the table.

(i) Which concentration of sodium thiosulphate gave the **slowest** reaction?
Choose **A**, **B**, **C**, or **D**. (1)

(ii) Which is the **most concentrated** solution of sodium thiosulphate?
Choose **A**, **B**, **C**, or **D**.
Explain your answer. (2)

(b) Changing the concentration changes the speed of the reaction.
Write about other ways of speeding up this reaction. (2)

(OCR Suffolk 1999)

Enzymes

6 This question is about catalase, an **enzyme** in vegetables.

Catalase acts as a catalyst for the splitting up of hydrogen peroxide.

hydrogen peroxide → water + oxygen

(a) Sam does an experiment at 25°C.

She uses 25 cm^3 of hydrogen peroxide solution and 1 cm^3 of catalase solution.

She measures the volume of gas given off each minute for five minutes.

Here are her results.

Time in minutes	0	1	2	3	4	5
Volume of gas in cm^3	0	25	40	48	50	50

(i) Plot the points on some graph paper. (2)

(ii) Finish the graph by drawing the best line through the points. (1)

(iii) Sam does the experiment again but this time at 30°C.

Draw, on the same grid, the graph she would expect to get. (2)

(b) Enzymes are used when beer is made by fermentation.

Enzymes in yeast act on sugar solution.

Finish the word equation for fermentation by choosing words from this list.

carbon dioxide

ethanol

oxygen

sugar solution

water

Sugar solution $\xrightarrow{\text{yeast}}$ .... + ...

.... (2)

(OCR Nuffield 1999)

7 A wine-maker dissolved sugar, yeast and grape juice in water. The mixture was added to a large clean bottle. The heater and airlock were fitted. The mixture fermented forming alcohol.

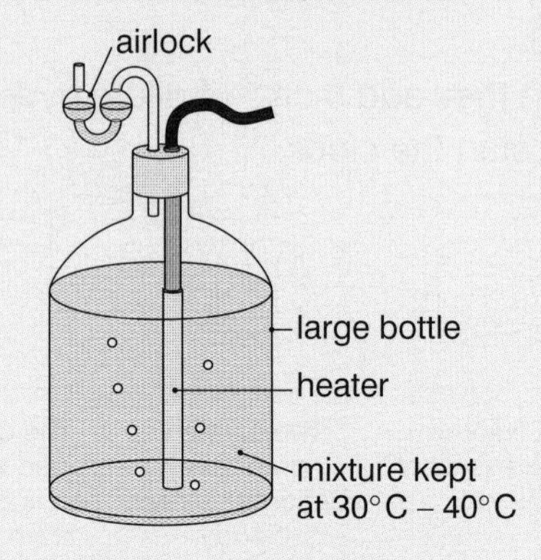

(a) What was the purpose of the yeast? (1)

(b) Name the gas formed in the process. (1)

(c) Explain why it was important to keep the temperature of the mixture at 30°C–40°C. (2)

(d) Explain why an airlock was used. (2)

(e) The bubbles of gas coming through the airlock were counted for one minute each day. The results are shown on the graph.

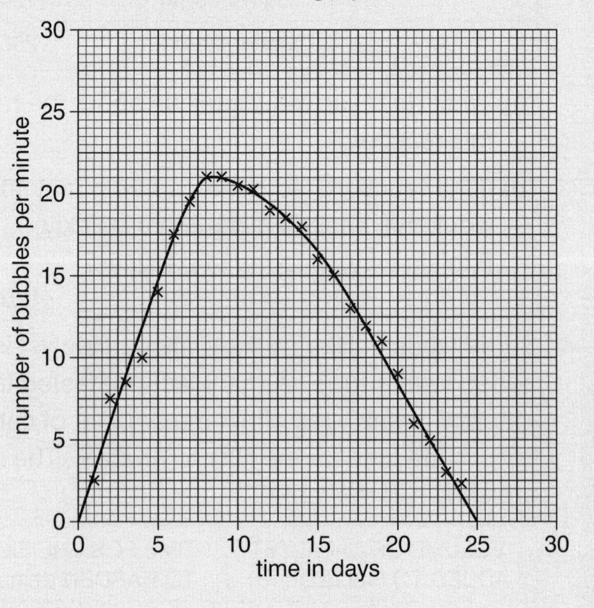

(i) When was the rate of formation of alcohol greatest? (2)

(ii) When did the formation of alcohol stop? (1)

(iii) Give **two** reasons why the formation of alcohol stopped. (2)

(AQA SEG 1999)

More about reactions

8 The temperature was taken before and after four chemical reactions.

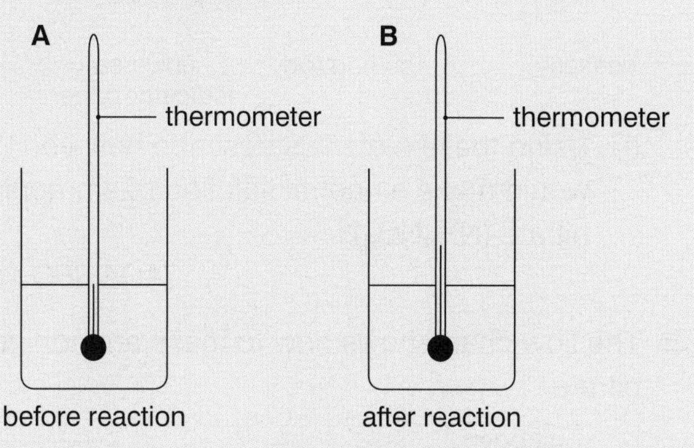

Reaction	*Thermometer* **A** *Reading* (°C)	*Thermometer* **B** *Reading* (°C)
P	18	17
R	19	17
S	20	25
T	19	25

(i) In which reaction, **P**, **R**, **S** or **T**, is **most** energy given out? (1)

(ii) What is the word used to describe reactions in which energy is given out? (1)

(WJEC)

9 Look at the diagram. It shows a different experiment.

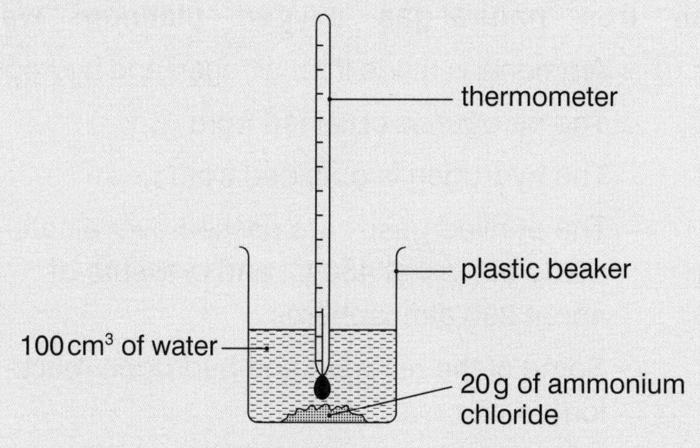

Jo measures the temperature of the water.

She adds 20 g of ammonium chloride to the water.

Jo stirs the water until the ammonium chloride dissolves.

She then measures the temperature again.

(i) Look at the results table.

temperature of the water at start in °C	20
temperature of the solution at the end in °C	15
temperature change in °C	

Which number is missing from the table? (1)

(ii) Is the process endothermic or exothermic? Explain your answer. (1)

(OCR Suffolk 1999 Reserve Q)

10 Mountaineers can warm their food in self-heating, sealed containers.

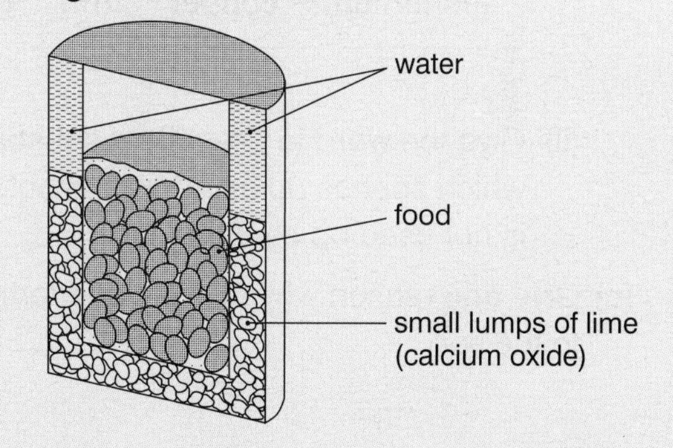

(a) The water is allowed to react with the lime. The heat from the reaction warms the food. What type of reaction causes a rise in temperature? (1)

(b) Some students investigated the effect of adding different sized lumps of lime to water. The results of their investigation are shown.

Time in minutes	**Temperature in °C**		
	Large lumps of lime	**Small lumps of lime**	**Powdered lime**
0	18	18	18
1	19	20	28
2	21	23	43
3	24	27	63
4	28	32	88
5	33	38	100

What do these results show? Give an explanation for your answer. (2)

(c) suggest and explain **one** disadvantage of using powdered lime to heat food. (2)

(AQA SEG 2000)

Further questions on Chemical reactions and their useful products

Fertilisers

11 Ammonia, NH_3, is an important chemical which is used to make fertilisers.

(a) The industrial process for making ammonia uses two gases.
Name the **two** gases which are reacted together to make ammonia.
...... and ...... (1)

(b) (i) Choose from the box below the metal used to speed up the manufacture of ammonia.

aluminium	**copper**	**iron**	**gold**

(1)

(ii) Give the word to describe a substance which speeds up a chemical reaction but is not used up by the reaction. (1)

(c) Give **one** reason why farmers add fertilisers to the soil. (1)

(WJEC)

12 As the world population increases there is a greater demand for fertilisers.

(a) The amount of nitrogen in a fertiliser is important.

(i) How many nitrogen atoms are there in the formula, NH_4NO_3? (1)

(ii) Work out the relative formula mass of ammonium nitrate, NH_4NO_3.

Relative atomic masses: H = 1; N = 14; O = 16. (1)

(b) Ammonium nitrate (NH_4NO_3) is manufactured by neutralising nitric acid (HNO_3) with ammonia (NH_3) solution.

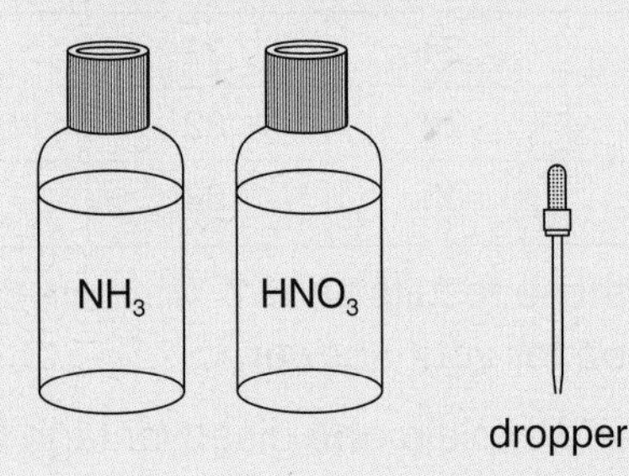

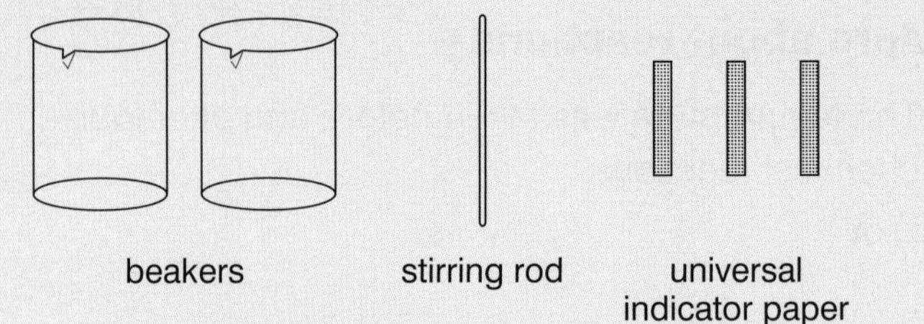

(i) Using these materials describe how you would make a neutral solution of ammonium nitrate (NH_4NO_3). (4)

(AQA SEG 2000)

13 The flow chart shows how to make ammonium nitrate.

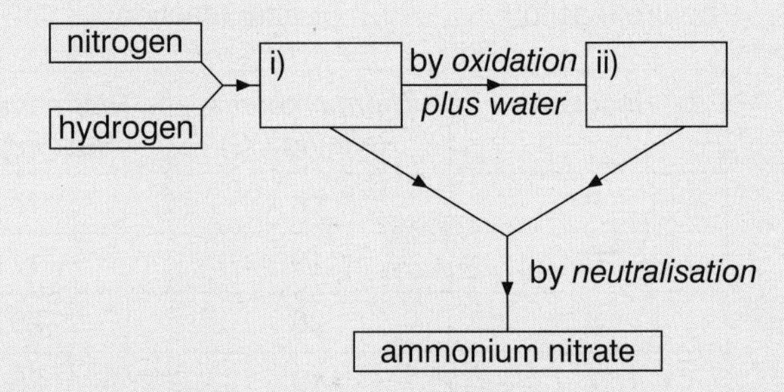

(a) Complete the chart by naming i) and ii) in the two boxes. (2)

(b) Explain why many farmers use ammonium nitrate. (2)

(AQA 2001)

14 Ammonia is manufactured by the Haber process.

(a) Choose words from the list to complete the passage.

air iron natural gas oxygen platinum water

Ammonia is made from nitrogen and hydrogen.

The nitrogen is obtained from ......

The hydrogen is obtained from ......

The purified gases are passed over a catalyst made of ...... at 450°C, and pressure of about 200 atmospheres.

Some of the nitrogen and hydrogen reacts to form ammonia. (3)

(b) (i) Calculate the relative formula mass (M_r) of ammonia, NH_3.

(Relative atomic masses: H = 1, N = 14)

(ii) Calculate the percentage of nitrogen in ammonia. (3)

(AQA 2000)

Section Four

Chemical patterns and bonding

In this section you will find out more what we find inside atoms and how this helps us to make sense of the world. You will learn about how atoms bond to each other and about the Periodic Table.

CHAPTER 14 Inside ATOMS

14a Protons, neutrons and electrons

Have you heard of the term 'splitting the atom'?
200 years ago scientists didn't think this was possible.

It was then that **John Dalton** put forward his ideas about atoms. Although the ancient Greeks were the first to think about atoms, science around 1800 became more firmly based on experiments. Dalton explained his observations by saying that atoms were like tiny snooker balls. They were so small that nobody could see them.
They were solid and couldn't be split into anything simpler.
Each chemical element was made of its own type of atom.

John Dalton (1766–1844) was born in Cumbria. For most of his life he taught in Manchester.

a) Look at Dalton's idea about chemical elements above.
Do we still use his idea? (See page 11.)

However, we now know that atoms can be split.
There are actually particles that are smaller than atoms.
Imagine if all the 100 or so chemical elements had atoms that each contained different types of particle.
Science would be very complicated.
Fortunately for us, there are only 3 main types of particle inside any atom (called sub-atomic particles).
These are **protons, neutrons and electrons**.

b) If there are only 3 types of particle inside any atom, how do you think we can have so many different atoms?

We now think that any atom is made up of:

- a tiny, dense core that is positively charged.
  This is called the **nucleus**.
- even smaller negatively charged particles that orbit around the nucleus. These are called **electrons**.

Look at the diagram opposite:

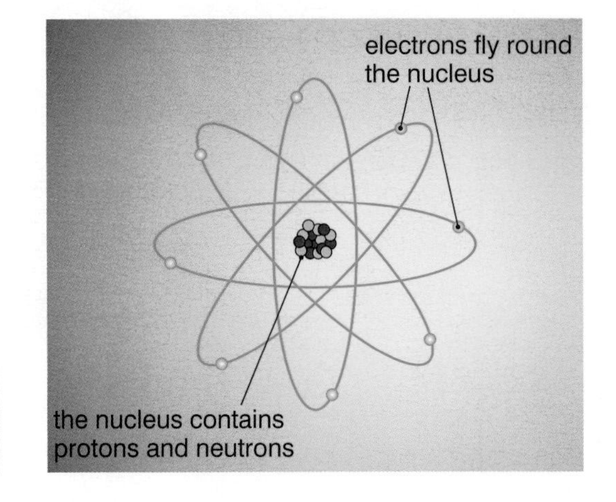

c) What charge must be on the nucleus to stop the negatively charged electrons flying out of an atom?

Inside the nucleus

Remember that the nucleus is the centre of an atom.

> The nucleus contains 2 types of particle:
> **protons and neutrons**.

These are the heavy particles in an atom.
The protons and neutrons have the same mass as each other.
However **p**rotons are **p**ositively charged whereas
neutrons are **neutr**al.

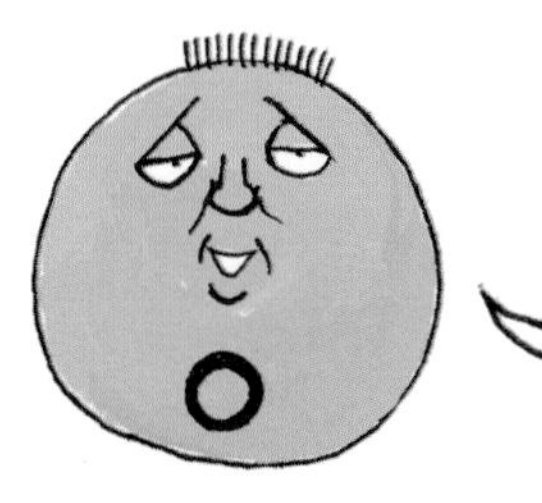

Electrons

The electrons that zoom around the nucleus
are really tiny. In fact, it would take about 2000
of them to have the same mass as a proton or neutron.
Remember that they are negatively charged.

Remind yourself!

1 Copy and complete:

There are …… types of sub-atomic particles found in atoms: ……, neutrons and ……

The …… and …… are found in the centre (or ……) of the atom. The …… whizz around the outside of it.

2 Do some research and write a report on:

a) The ideas of the ancient Greek Democritus about atoms.

b) The work of the English scientist John Dalton.

14b Structure of the atom

Although there are only 3 types of sub-atomic particle to learn about, it is tricky to remember everything. Here is a table to help you:
The charges and masses are given relative to each other.

Sub-atomic particle	Mass	Charge
proton	1	+1
neutron	1	0
electron	0 (almost)	–1

a) You can add up the mass of sub-atomic particles to find the mass of an atom.
Why can we ignore the mass of the electrons in the sum?

Arrangement of electrons

As you know, the electrons fly around the nucleus. Scientists have found out that they orbit the nucleus in **shells** (sometimes called **energy levels**).
These shells spread out from the centre of an atom.

Electrons fill up the shells from the middle outwards.
So the shell nearest the nucleus fills first.
Then new shells are filled.

Each shell can only hold a set number of electrons before filling up.

Remember that:

The 1st shell can hold **2** electrons.
The 2nd shell can hold **8** electrons.
The 3rd shell can hold **8** electrons.

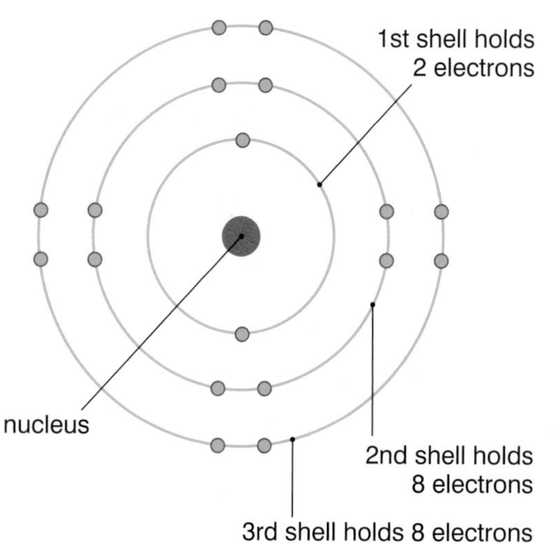

b) How many electrons can fit in the first shell (which is sometimes called the lowest energy level)?

Look at the examples of atoms below:

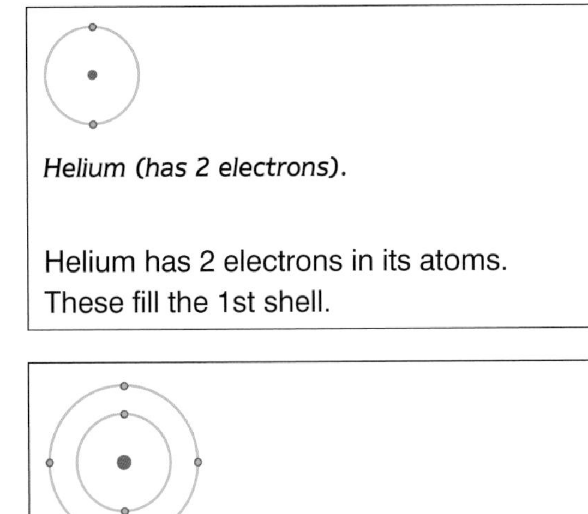

Helium (has 2 electrons).

Helium has 2 electrons in its atoms.
These fill the 1st shell.

Carbon (has 6 electrons).

This carbon atom has 6 electrons.
The first 2 go into the 1st shell.
That leaves 4 electrons to occupy the 2nd shell.

Here is a sodium atom:

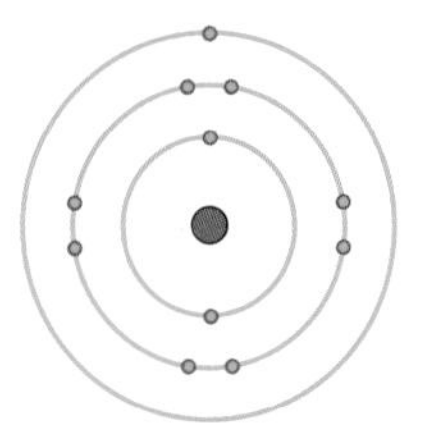

Sodium (has 11 electrons).

c) Explain how the electrons fill up the shells in a sodium atom?

Remind yourself!

1 Copy and complete:

If a ...... has a charge of +1 and a mass of 1, then a ...... has a charge of 0 and a mass of ......

The electrons are found in ...... around the ...... The 1st shell can hold up to ...... electrons, and the 2nd and 3rd shells up to ...... electrons.

2 Draw diagrams of these atoms to show how their electrons are arranged:

a) hydrogen (which has 1 electron)

b) boron (which has 5 electrons)

c) neon (which has 10 electrons)

d) magnesium (which has 12 electrons)

e) chlorine (which has 17 electrons)

f) calcium (which has 20 electrons).

14c Atomic short-hand

How did you like drawing all those circles in the last question on page 171?
Chemists have made up a way to get around all that.
They show the number of electrons in each shell by numbers, starting from the 1st shell and working outwards.
They can show these **electronic structures** like this:

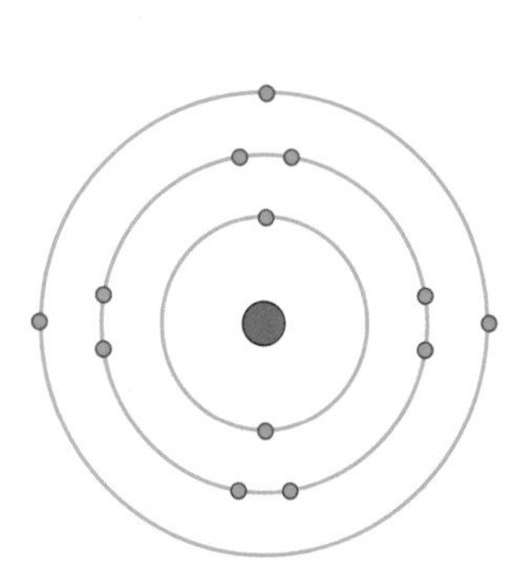

Aluminium has an electronic structure of ***2, 8, 3***

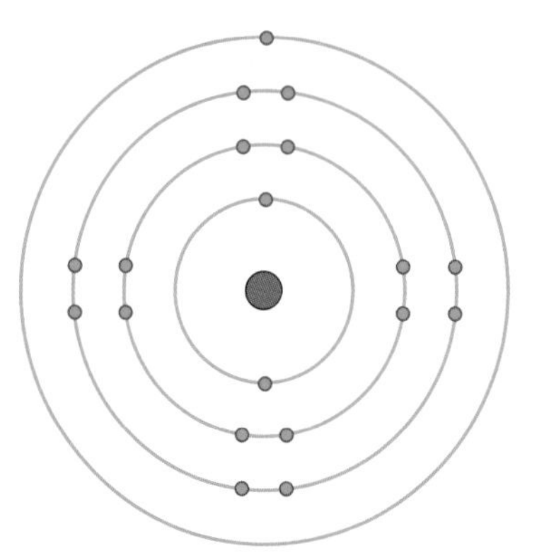

Potassium has an electronic structure of ***2, 8, 8, 1***

a) What do you notice when you add up all the numbers in the electronic structure? What does it equal?

b) Write down the electronic structure of a fluorine atom (which has 9 electrons).

Chemists also have a short-hand way of showing the numbers of protons, neutrons and electrons in an atom.

Atomic number

The **atomic number** tells us the number of protons in an atom.
In fact, it is sometimes called the proton number.
All atoms are neutral. They carry no overall charge.
So this number also tells us the number of electrons (because the positive protons must be cancelled out by the negative electrons).

Atomic number = the number of protons (which equals the number of electrons)

This is shown in chemical short-hand like this:

$${}_{3}\mathrm{Li}$$

This tells us that the atomic number of lithium is 3.

c) How many protons are there in an atom of lithium?

d) How many electrons are there in a lithium atom?

Mass number

Notice that the atomic number doesn't tell us anything about the number of neutrons in an atom.
We have to work this out from the **mass number**.

Mass number = the number of protons + neutrons

We show the mass number like this:

^{7}Li

This means that this lithium atom has a total of 7 protons and neutrons in its nucleus.

e) Why is this called the ***mass*** number?

But how many of the 7 particles in lithium's nucleus are neutrons?
To answer this we need its atomic number as well.
We know the atomic number of lithium is 3.
Therefore it has 3 protons.
So there must be 4 neutrons to make up its mass number of 7.
You can remember this with an equation:

Number of neutrons = mass number − atomic number

All the information we need to know about an atom can be shown like this:

Mass number 7
Atomic number 3 $^{7}_{3}Li$

The electronic structure gives us the complete picture:

2, 1

f) Draw a labelled diagram to show all the information about the Li atom.

Remind yourself!

1 Copy and complete:

The ...... structure of an atom can be shown like this: 2, 8, ... , 1.

The number of protons in an atom of an element is given by its ...... number. This also tells us the number of ......

The ...... number of an atom gives us the number of protons plus ...... in the ...... of the atom.

2 Write down the electronic structures of these atoms:

a) beryllium (atomic number 4)

b) sulphur (atomic number 16)

c) calcium (atomic number 20)

3 Give all the information you can about these atoms:

a) $^{14}_{7}N$ b) $^{31}_{15}P$ c) $^{40}_{18}Ar$

▶▶▶ 14d Isotopes

Have you heard of the word 'isotope' before?
You might have come across it when people talk about treating cancer or pollution from radioactive waste.
But what is an isotope?

There are some elements that are made up of atoms that have different numbers of neutrons.
These are called **isotopes**.
They are atoms of the same element because they have the same number of protons.
Every chemical element has its own unique number of protons. For example, any atom with 1 proton must be an atom of hydrogen. Most hydrogen atoms have no neutrons at all. So their atomic number is 1 and their mass number is 1 ($^{1}_{1}H$). Look opposite:

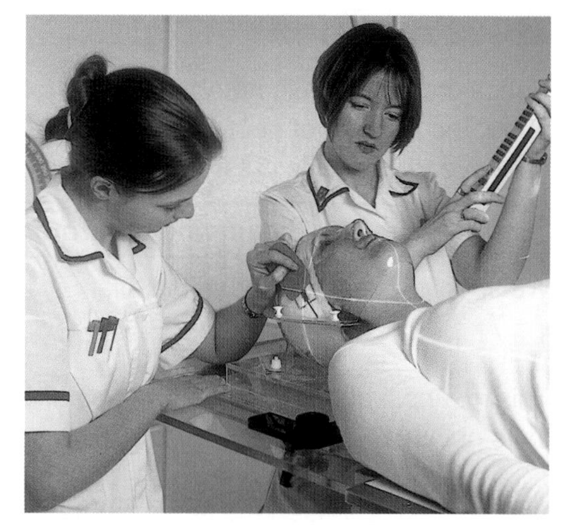

Radioactive isotopes are used to kill cancer cells.

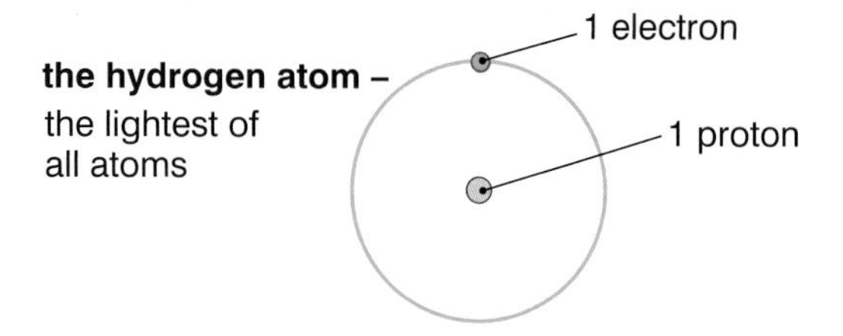

a) How many electrons does every atom of hydrogen have?

But there are two other hydrogen atoms that do have neutrons.
$^{2}_{1}H$ has 1 neutron, and $^{3}_{1}H$ has 2 neutrons.
Look at the 3 isotopes of hydrogen below:

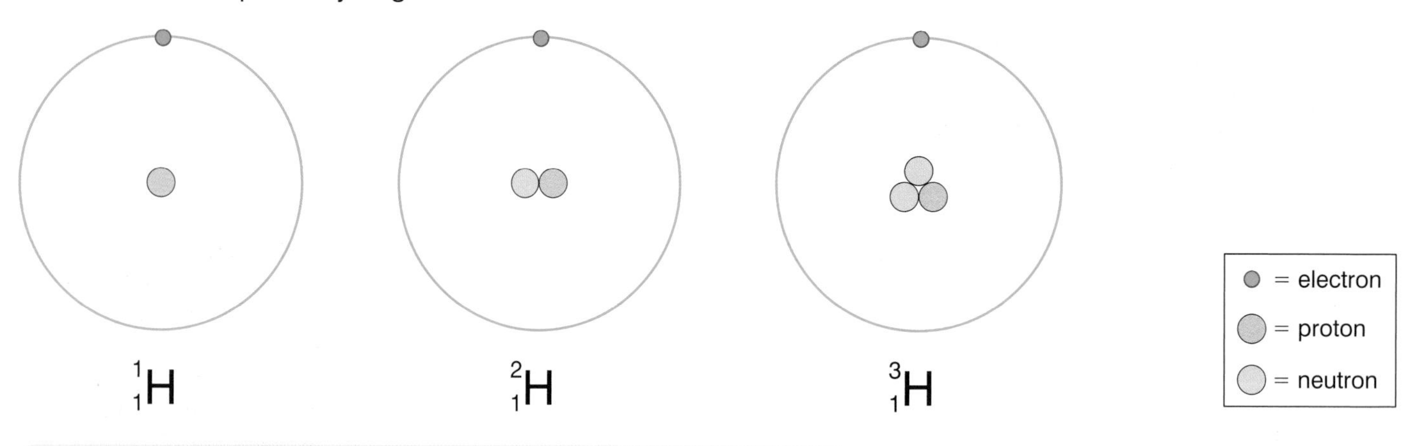

b) How does an atom of $^{1}_{1}H$ differ from the isotope $^{3}_{1}H$?

c) What can you say about the atomic numbers and mass numbers of the isotopes above?

So we can think of **isotopes** as:
- atoms of the same element with different numbers of neutrons

or
- atoms with the same atomic number but different mass numbers

or
- atoms with the same number of protons but different numbers of neutrons.

All these definitions of isotope mean exactly the same thing!

Here is another example:

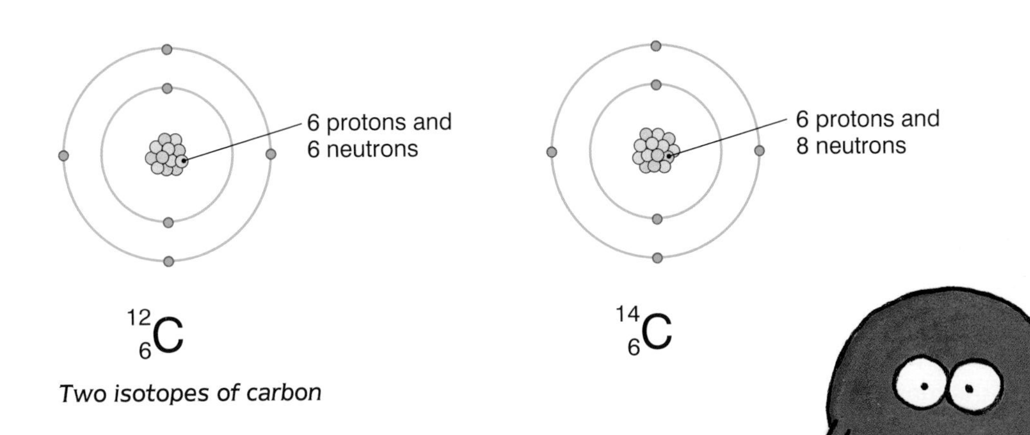

Two isotopes of carbon

These are isotopes of the element carbon.
They are identical in every way except that one isotope is heavier than the other.

Isotopic twins!

d) Which isotope of carbon shown above is heavier? Why is this isotope heavier?

e) Which of the two isotopes has more electrons?

The reactions of isotopes of an element will be the same.
That's because it is the number of electrons that are important in chemical reactions.
(You can find out more about this in the next chapter.)
So both the isotopes above will react with oxygen when heated, and both will form carbon dioxide.

Sometimes we show isotopes by the name of the element followed by its mass number.
So the isotopes we have looked at so far are:
hydrogen-1, hydrogen-2 and hydrogen-3
as well as
carbon-12 and carbon-14.

Not all isotopes are radioactive, but radioactive isotopes of iodine are still causing health problems years after this accident in a nuclear power station at Chernobyl in 1986.

Remind yourself!

1 Copy and complete:

The atoms of isotopes of an ...... contain equal numbers of ...... and ......, but different numbers of ......

This means that they have the same ...... number but different ...... numbers.

They will have exactly the same chemical ......

2 a) Complete this table to show the numbers of protons, neutrons and electrons in these isotopes of magnesium:

Isotope	Protons	Neutrons	Electrons
$^{24}_{12}Mg$			
$^{25}_{12}Mg$			
$^{26}_{12}Mg$			

b) How do the isotopes differ?

Summary

Inside atoms we find **protons, neutrons and electrons**.
Here are their properties:

Sub-atomic particle	Mass	Charge
proton	1	+1
neutron	1	0
electron	0 (almost)	−1

The protons and neutrons are in the centre of the atom called the **nucleus**. This is where the mass is concentrated.

The electrons whizz around the nucleus in **shells** (or **energy levels**).
The 1st shell holds up to **2** electrons.
The 2nd shell can hold up to **8** electrons.
The 3rd shell holds **8** electrons.

We can show an atom's **electronic structure** quickly using numbers e.g. 2, 8, 8, 1 for the potassium atom below:

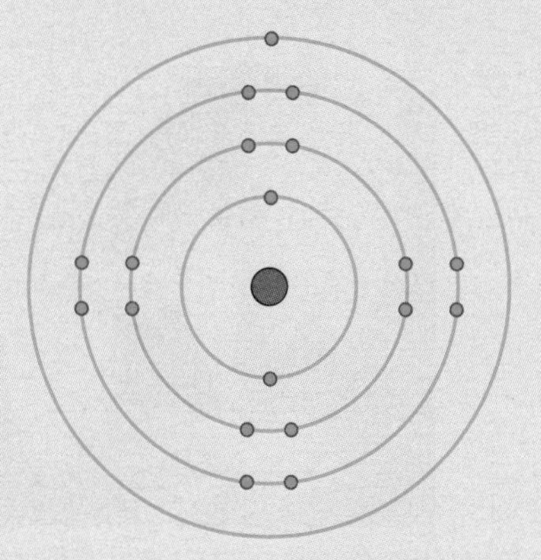

The potassium atom has 2 electrons in its first shell, 8 electrons in its second shell, 8 electrons in its third shell, and 1 electron in its fourth (outermost) shell.

The **atomic number** tells us how many protons (and therefore electrons) there are in an atom.

The **mass number** tells us the number of protons + neutrons.

(Number of neutrons = mass number – atomic number)

We can show these like this:

mass number → $^{14}_{7}N$
atomic number →

Isotopes of an element contain different numbers of neutrons.

Questions

1 Copy and complete:

The ...... of an atom is called its nucleus.

In the nucleus we find ...... and ......
The ...... zoom around the nucleus in ......

An atom whose ...... structure is 2, 6 has ...... electrons in the 1st ...... and ...... electrons in the 2nd ......

The atomic number of an atom gives us the number of ...... which equals the number of ...... in an atom.

The ...... number tells us the total number of ...... and ...... in an atom.

Isotopes have the same ...... number but different ...... numbers.

2 How many protons, neutrons and electrons are there in these atoms:

a) hydrogen
(atomic number = 1, mass number = 1)

b) oxygen
(atomic number = 8, mass number = 16)

c) neon
(atomic number = 10, mass number = 20)

d) potassium
(atomic number = 19, mass number = 39)

e) zinc
(atomic number = 30, mass number = 65)

3 Write down the electronic structures of the following atoms:

a) helium (atomic number 2)

b) boron (atomic number 5)

c) fluorine (atomic number 9)

d) sodium (atomic number 11)

e) aluminium (atomic number 13)

f) argon (atomic number 18)

g) calcium (atomic number 20)

4 How many protons, neutrons and electrons do the following atoms have?

Also write down the electronic structure of each atom in a table like this:.

Atom	Protons	Neutrons	Electron	Electronic structures
$^{9}_{4}Be$				
$^{27}_{13}Al$				
$^{28}_{14}Si$				
$^{31}_{15}P$				

5 Here is an atom of boron:

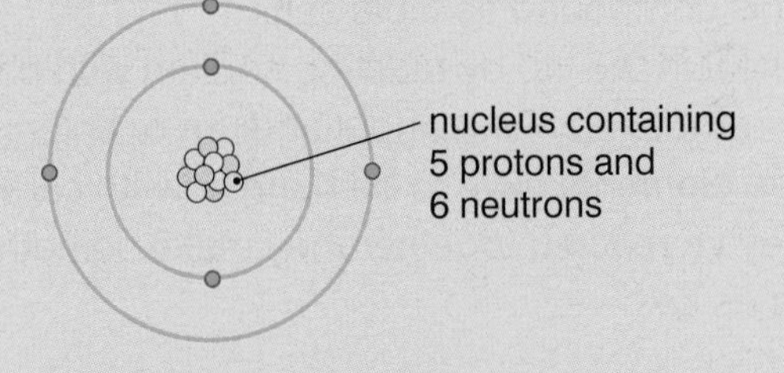

a) Show the information given above about a boron (B) atom using short-hand as in the table in question 4.

b) How many protons, neutrons and electrons are there in $^{119}_{50}Sn$?

6 Here are two atoms of the element uranium:

$^{235}_{92}U$ and $^{238}_{92}U$

a) What is the difference in the numbers of protons, neutrons and electrons in the two atoms?

b) What are these two atoms examples of?

c) Uranium reacts with fluorine to make UF_6. Which atom would react more quickly?

d) Write a short report on the use of uranium in a nuclear reactor.

CHAPTER 15 BONDING

15a Losing and gaining electrons

Have you ever thought what happens to atoms
when they react with each other?
Now that you know about what's inside an atom,
you can understand how atoms bond to each other.

Sodium and chlorine were pretty wild when they were single. Now they've 'tied the bond', they're totally different!

Think of the example of sodium chloride:
You know this compound as table salt.
We can make it in the lab by reacting sodium with chlorine.
Can you remember anything about sodium or chlorine?
Both of these elements have to be handled with care.
Yet once they've reacted together they make something
as harmless as salt:

sodium + chlorine ⇨ sodium chloride

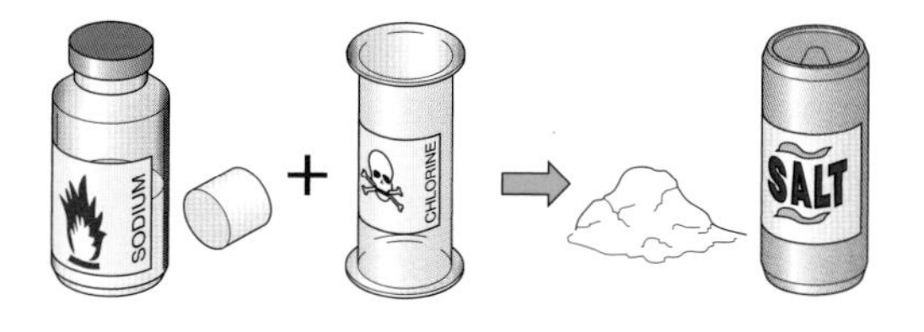

a) What is the metal sodium like?
b) What is chlorine gas like?
c) Describe sodium chloride.
d) What can you say about the reactants and products in a chemical reaction?

Let's look at the atoms of sodium and chlorine that we start with:

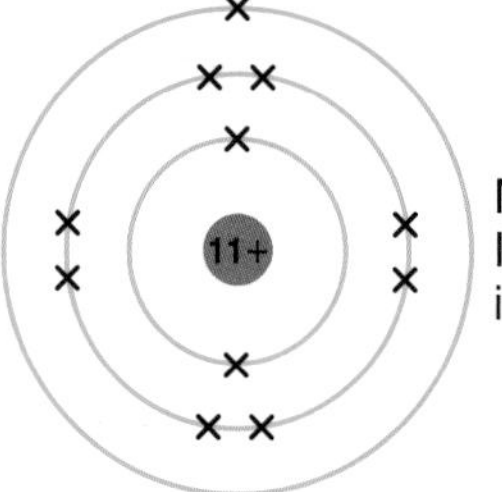

Na atom.
It has 1 electron
in its outer shell

A sodium atom has 11 electrons (2, 8, 1).

17+

Cl atom.
It has 7 electrons
in its outer shell

A chlorine atom has 17 electrons (2, 8, 7).

The most stable atoms have a full outer shell of electrons.

e) How many electrons does a sodium atom have in its outer shell?

f) How many electrons does a chlorine atom have in its outer shell?

To Chlorine

Roses are red
Violets are blue
Accept this electron
From me to you!

luv Sodium
x

Sodium would have a full outer shell if it could get rid of 1 electron.
Chlorine would have a full outer shell if it could gain 1 electron.

Can you see what happens to the electrons when sodium and chlorine react together?
You've probably guessed!
Sodium gives its outer electron to chlorine.
Then they both have full outer shells (and everyone's happy!).

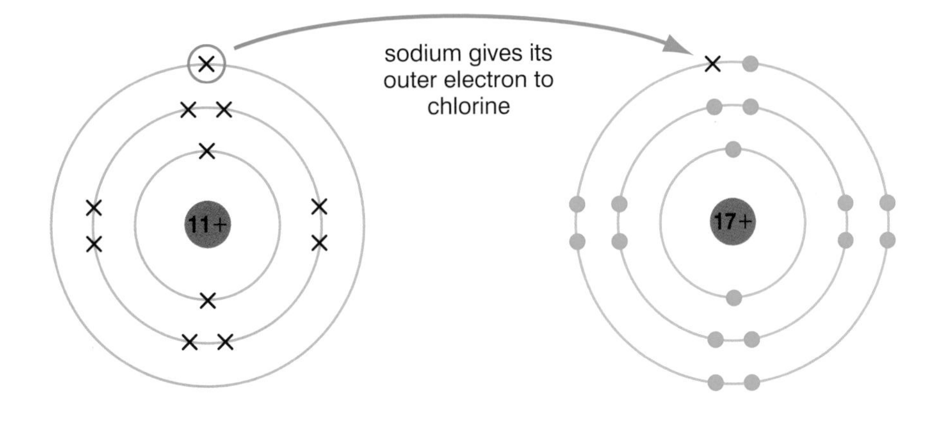

When the atoms lose or gain electrons they become charged.
Remember that electrons are negative and sodium has lost an electron. But it still has 11 positive protons in its nucleus.
So when you count up the charge we get:
10 electrons (10–) and 11 protons (11+) which equals 1+.
Sodium forms what we call a **sodium ion, Na^+**.

The opposite happens to chlorine.
Once it has reacted, it has an extra electron.
It now has 18 electrons (18–) and 17 protons (17+):
18– and 17+ which equals 1–.
So chlorine forms a **chloride ion, Cl^-**.

Remind yourself!

1 Copy and complete:

Atoms can transfer ...... to each other when they ...... together. For example, sodium ...... its ...... electron to chlorine.

This makes the sodium a ...... ion (with the formula ......) and chlorine forms a ...... chloride ion (whose formula is ......)

2 Explain what happens to the electrons in each atom when the following atoms react with each other:

a) lithium (with 3 electrons) and chlorine (with 17 electrons)

b) sodium (with 11 electrons) and fluorine (with 9 electrons).

15b Ionic bonding

On the previous page, we saw how a sodium atom transfers an electron to a chlorine atom when they react.
The atoms then become charged particles called **ions**.

We can show how the ions form like this:

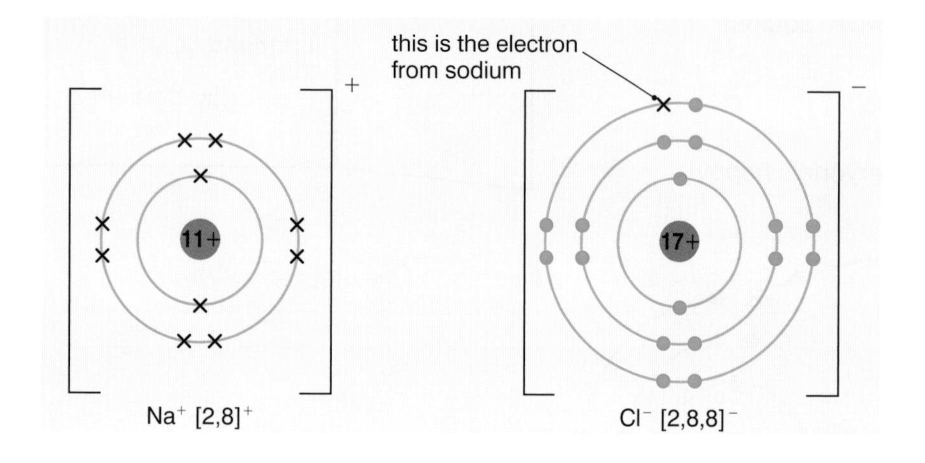

You know that opposite charges attract.
The positive sodium ion and the negative chloride ion are strongly attracted to each other. This attraction of the oppositely charged ions sticks them to each other. It is called an **ionic bond**.

Opposites attract!

The attraction between oppositely charged ions is called ionic bonding.

The compound formed is called an ionic compound.

Ionic compounds are formed when metals react with non-metals.

When metal meets non-metal, the sparks they do fly.
'Give me electrons!' non-metals do cry.
'You've got a deal,' the metal responds,
Then ions attract to form their bonds.

Sodium reacts with chlorine to form the ionic compound, sodium chloride.

a) Which of these compounds would you expect to have ionic bonding? Explain your reasoning.
sulphur dioxide
iron chloride
hydrogen sulphide
magnesium bromide

Giant ionic lattice

In the actual reaction between sodium and chlorine, millions and millions of ions form and line up next to each other in giant structures.
The structure is called a **giant ionic lattice**.

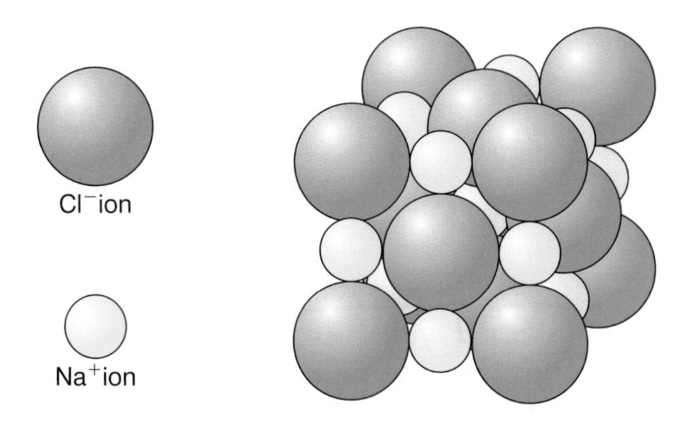

Part of the giant ionic lattice of sodium chloride.

The forces of attraction between all these oppositely charged ions makes it very difficult to melt an ionic compound.

b) What happens to the particles in a solid when it melts? (See page 41.)

Ionic compounds have high melting points and high boiling points.

Once they are melted, the ***ions can move around***.
They can also move around when they are dissolved in water.
(See page 41.)
This explains why:

Ionic compounds don't conduct electricity when solid, but do when molten or dissolved in water.

Remind yourself!

1 Copy and complete:

When a metal reacts with a ......, millions of positive and ...... ions line up to form a giant ...... lattice.

Ionic compounds have ...... melting points.

They do not ...... electricity when they are ......, but do when they are ...... or ...... in water.

2 a) Magnesium's atomic number is 12. What is its electronic structure?

b) How many electrons does it lose when it forms an ion? What will its charge be?

c) Oxygen's atomic number is 8. What is its electronic structure?

d) How many electrons does it gain when it forms an ion? What will its charge be?

e) Explain, using diagrams, how magnesium oxide is formed.

15c Sharing electrons

So far we have seen how metals bond to non-metals.
Metals give electrons to non-metals to form ions.
But what happens in compounds like water, H_2O?
Hydrogen and oxygen are both non-metals.
They both need to gain electrons.

It's like you and a friend who both want the same magazine
but there's only one left. What do you do?
No, not fight it out! You share the magazine.
And that's what atoms of non-metals, like hydrogen and oxygen, do.

> When atoms bond together by ***sharing electrons***, we call it **covalent bonding**.

Look at the atoms of hydrogen and fluorine below:

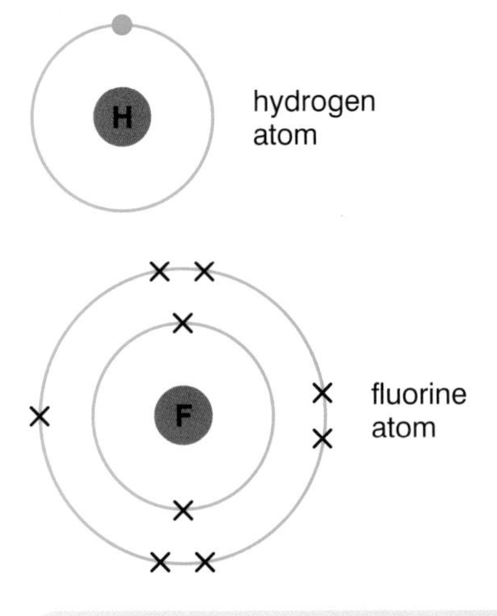

a) How many electrons does the 1st shell hold?
b) How many electrons does the hydrogen atom need to gain to fill its first shell?
c) How many electrons does the 2nd shell hold?
d) How many electrons does the fluorine atom need to gain to fill its 2nd shell?

Both atoms can gain full outer shells by overlapping and sharing a pair of electrons. One electron in the pair comes from each atom.

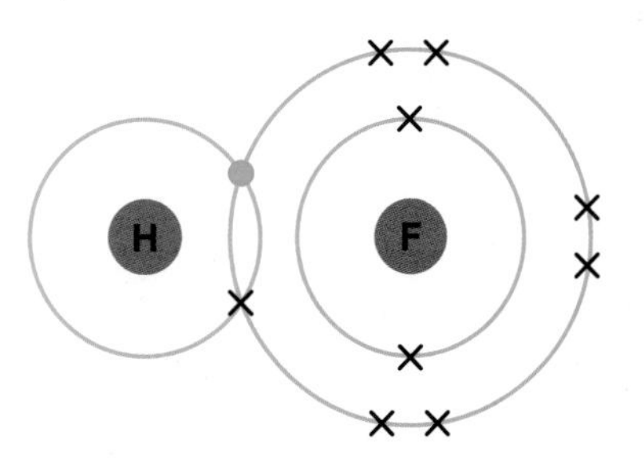

Atoms form a covalent bond *by overlapping* their outer shells and *sharing a pair of electrons*.

We can show these covalent bonds in different ways:

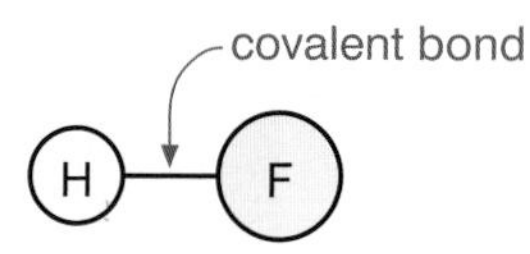

This is sometimes called a 'ball-and-stick' diagram.

Or we can just show the outer electrons, without drawing in the circles for the shells:

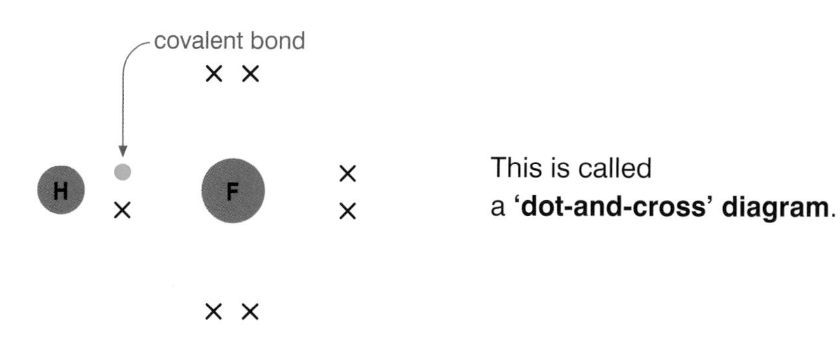

This is called a **'dot-and-cross' diagram**.

e) Why do you think that people use 'dot and cross' diagrams rather than drawing out each atom in full?

Here is another molecule whose atoms are joined to each other by a covalent bond:

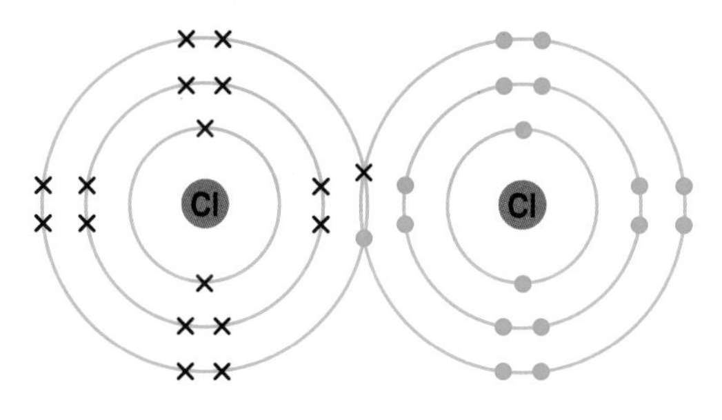

Remind yourself!

1 Copy and complete:

Covalent bonds form between atoms of non-……
These atoms both need to …… electrons. They do this by …… their outer shells.
A …… of electrons is …… in a covalent bond.
We can show the bonding in a '…… and cross' diagram.

2 a) Show the chlorine molecule above as:
i) a 'ball and stick' diagram
ii) a 'dot and cross' diagram.

b) Draw a molecule of fluorine, F_2, showing all its electrons and shells.

c) Draw a molecule of HCl, showing all its electrons and shells.
(Atomic numbers: F = 9, H = 1, Cl = 17)

▶▶▶ 15d More covalent molecules

When you enjoy a nice cool glass of water, you are drinking covalently bonded molecules.

Look at an atom of oxygen below:

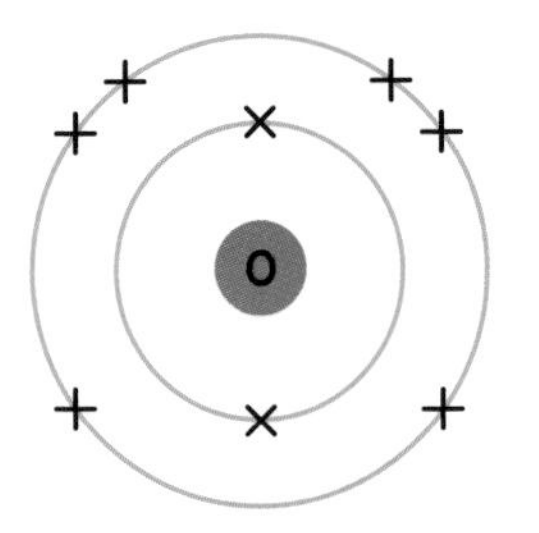

Water molecules contain covalent bonds.

a) How many electrons does oxygen need to gain a full outer shell?

b) Remember that hydrogen atoms need to gain one electron. So how many hydrogen atoms need to share with oxygen?

Look at the covalent bonds in a water molecule:

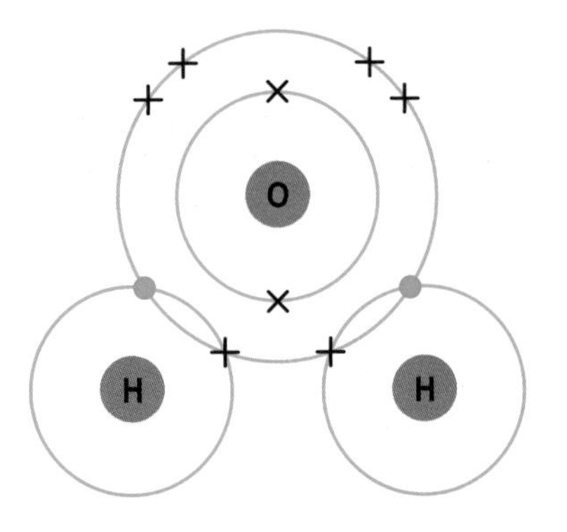

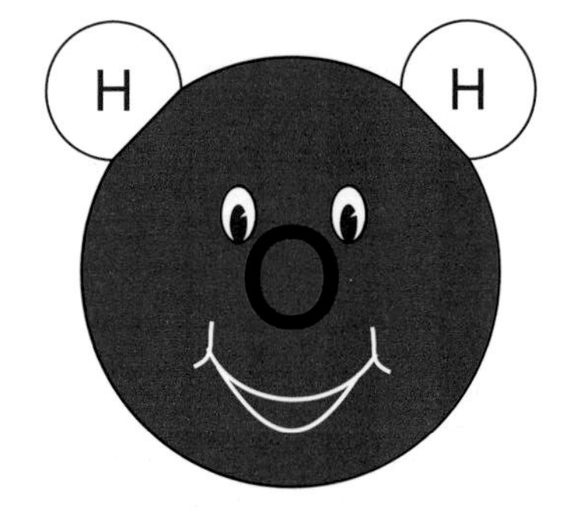

Can you see why the formula of water is H_2O (and ***not*** HO or H_3O etc.)?

Here are some other covalent molecules you might come across:

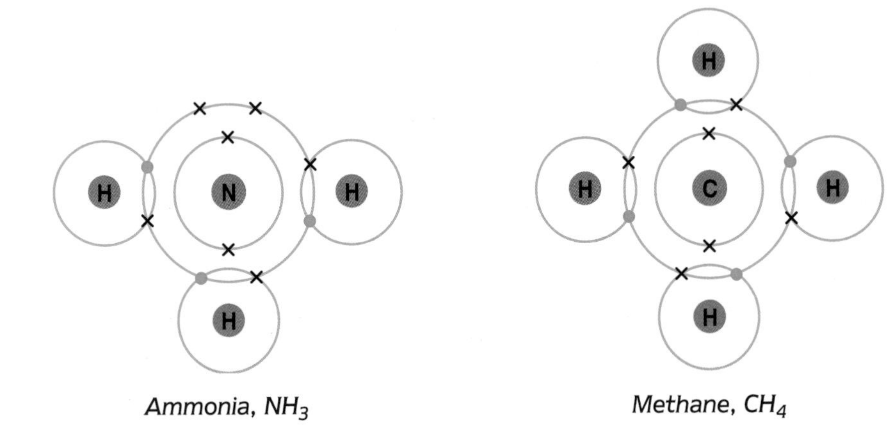

Ammonia, NH_3

Methane, CH_4

Properties of covalent substances

Let's look at the melting points and boiling points of the covalent molecules we have met so far:

Covalent molecule	Melting point (°C)	Boiling point (°C)
hydrogen fluoride (HF)	–83	20
chlorine (Cl_2)	–101	–35
water (H_2O)	0	100
ammonia (NH_3)	–78	–33
methane (CH_4)	–182	–164

These are all low values when you think of the ionic compounds we saw earlier.
For example, the melting point of magnesium oxide is over 2850°C!

c) Which substance in the table above has the lowest boiling point?
d) Which substance in the table above is an element?

weak forces **between** molecules

strong covalent bonds **within** each molecule

Methane has a simple molecular structure.

So we can say that:

Substances made of individual **molecules** have **low melting points** and **low boiling points**.

The covalent bonds themselves are strong, but the forces ***between*** individual molecules are only weak.
They are called molecular or **simple molecular** compounds.

Giant covalent structures

Some substances with covalent bonds do have high melting points. However, these are **giant structures** held together by millions of covalent bonds. They are not made of individual molecules.
Examples include diamond, graphite and silica (sand).

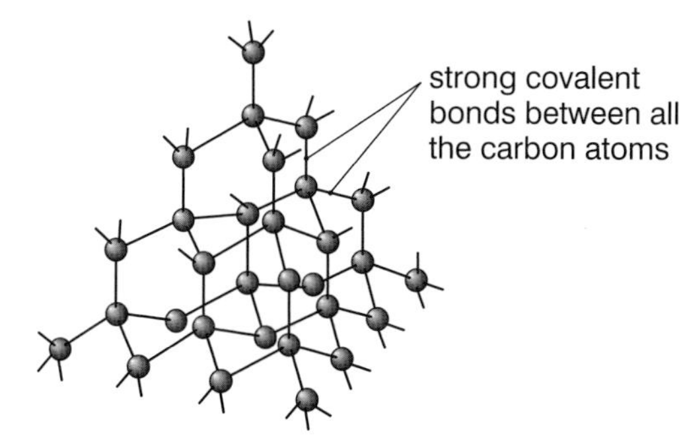

To melt diamond you would have to break millions of covalent bonds.

Remind yourself!

1 Copy and complete:

Substances made up of individual ……, such as H_2O, have …… melting points and …… points. Their …… bonds are strong within each ……, but there are relatively …… forces between them.

Substances with …… covalent structures, such as diamond or g……, have …… melting points.

2 Copy this table and put a tick in the correct column:

Substance	High melting point	Low melting point
ammonia		
sodium chloride		
methane		
oxygen		
magnesium oxide		

15e Metals

We have already looked at how important metals are in Chapter 2. We use them so much because they have some very useful properties.

Metals have many useful properties.

a) Look back to page 18 and list the general properties of metals.

Any model we have of how metal atoms bond to each other must be able to explain these properties.

Metallic bonding

You know from your work with ionic bonding that ***metal atoms tend to lose electrons***.
In ionic compounds they give away electrons to non-metal atoms.
But in an aluminium can there are only metal atoms.
So what happens then?

In metals, each atom gives electrons from its outer shell into a 'sea' of electrons. These electrons are then free to move around within the metal.
We can think of the electrons as 'sticking' the structure together.

Look at the diagram below:

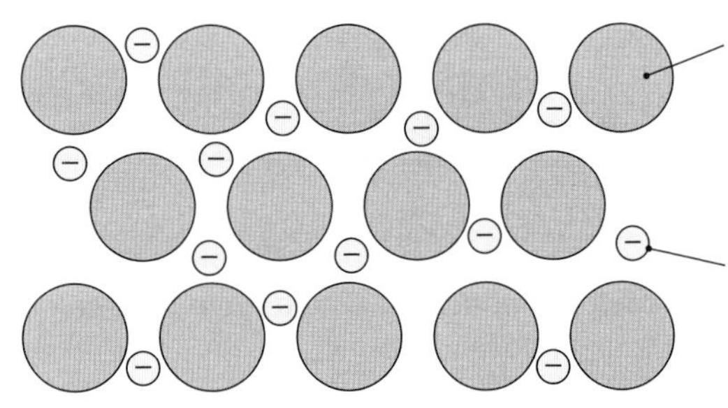

metal atoms (some people describe them as **positive ions** because they donate electrons into the 'sea' of electrons)

'sea' of electrons holds the metal atoms together

Metal atoms (or ions) are bonded to each other by free electrons.

b) Why do some people think of the particles in a metal as ions?

c) The atomic number of aluminium is 13.
How many electrons are in the outer shell of its atoms?

d) How many electrons does each aluminium atom donate into the 'sea' of electrons?

These free electrons are the reason why metals are good conductors of electricity. Look at the diagram below:

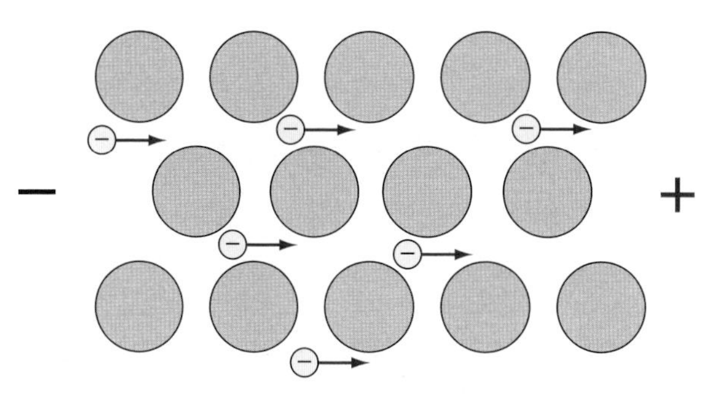

Electrons move towards the positive charge.

e) Explain what happens when a metal conducts electricity.

If you heat a metal at one end, its atoms vibrate faster at that end.
The vibration is passed down the metal.
But the free electrons can transfer the heat much faster as they move freely through the metal.
So metals are very good conductors of heat.

The metal atoms (or ions) are arranged in giant structures. Therefore, like other giant structures, they generally have high melting points:

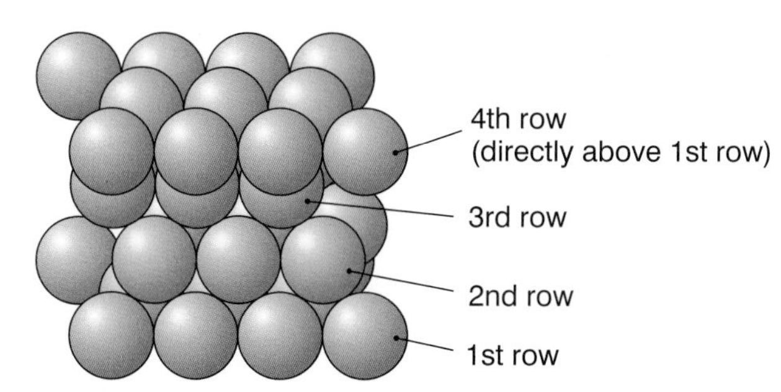

The regular structure of copper.

In metal structures, the gaps are few,
A 'sea' of electrons acts as a glue!
Lining up, row upon row,
On top of each other, the atoms do go.

If you strike the metal with a hammer, the layers of atoms (or ions) can ***slide over each other***. That's why metals can be 'worked' into different shapes without smashing.

Remind yourself!

1 Copy and complete:

Metal atoms (or ……) are bonded to each other by …… electrons. The electrons come from the …… shells of the metal ……

The atoms are arranged in g…… structures.

2 Explain why metals are:

a) good conductors of heat

b) malleable (can be bashed into different shapes with a hammer).

3 Make a list of 'odd' metals that don't have the general properties of metals.

Summary

Ionic bonding

Metals bond to non-metals in ionic compounds.
The metal atom gives one or more electrons to the non-metal atom.
This happens as the elements react together.
The charged particles formed are called **ions**.

Metal atoms form positive ions. (For example, Na^{+}, Mg^{2+}, Al^{3+}.)
Atoms of non-metals form negative ions. (For example, Cl^{-}, O^{2-}.)

The ions form regular structures called **giant ionic lattices**.
There are strong forces of attraction between oppositely charged ions.
So ionic compounds have high melting points.
They don't conduct electricity when they are solid,
but they do if you melt them or dissolve them in water.
The ions are then free to move around and carry the charge
through the liquid.

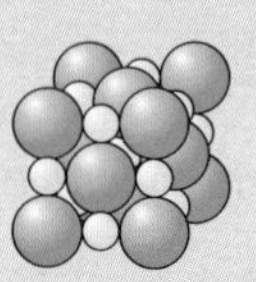

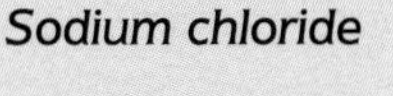

Sodium chloride

Covalent bonding

Atoms of non-metals can bond to each other by
sharing pairs of electrons. This is covalent bonding.

Many covalently bonded substance are made up of small
individual molecules. (For example, water.)
These have low melting points and low boiling points.

Other substances with covalent bonds have giant structures.
These have very high melting points. (For example, diamond.)

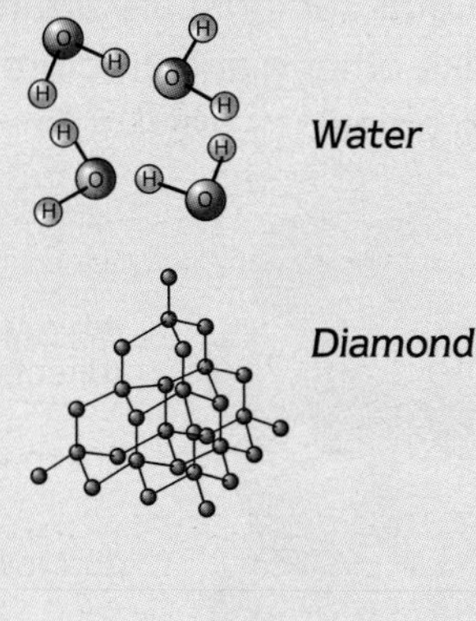

Water

Diamond

Metallic bonding

The atoms (or positive ions) in a metal are held to each other
by a 'sea' of free electrons. These electrons:

- hold the atoms (or ions) together in giant structures,
- can drift though the metal when it conducts,
- let the atoms (or ions) slip over each other when we
  hammer or stretch it.

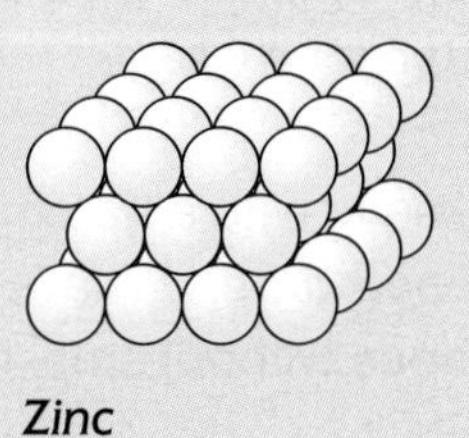

Zinc

Summary of different structures

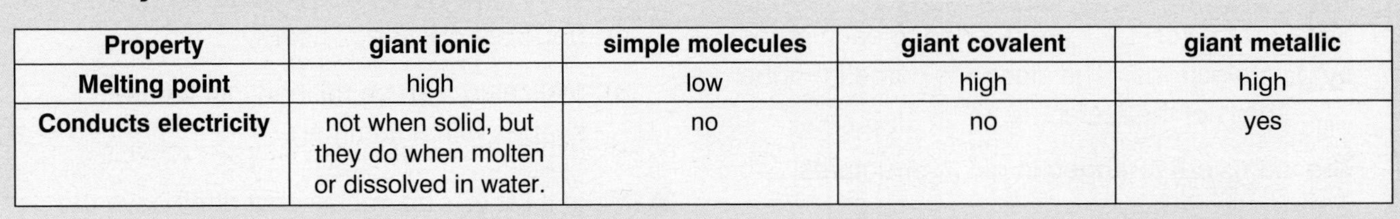

Property	giant ionic	simple molecules	giant covalent	giant metallic
Melting point	high	low	high	high
Conducts electricity	not when solid, but they do when molten or dissolved in water.	no	no	yes

Questions

1 Copy and complete:

The force of attraction between …… charged ions is called …… bonding.

The bond made when two …… share a …… of electrons is called a …… bond.

Metals are bonded by a 'sea' of …… electrons.

2 Show how the electrons are transferred between the following atoms:

a) Lithium (which has 3 electrons) and fluorine (which has 9 electrons).

b) Sodium (which has 11 electrons) and chlorine (which has 17 electrons).

c) Magnesium (which has 12 electrons) and oxygen (which has 8 electrons).

d) Calcium (which has 20 electrons) and 2 chlorine atoms (which each have 17 electrons).

3 What will the charge be on the ions formed by the following atoms:

a) lithium (atomic number 3)

b) sodium (atomic number 11)

c) potassium (atomic number 19)

d) aluminium (atomic number 13)

e) fluorine (atomic number 9)

f) chlorine (atomic number 17)

g) sulphur (atomic number 16).

4 Copy these diagrams and fill in the electrons in the outer shells :

a) hydrogen chloride (HCl)

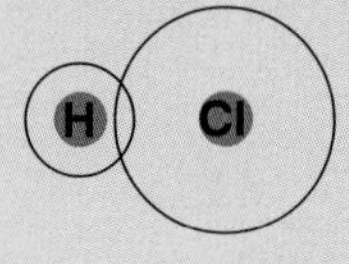

b) water (H_2O)

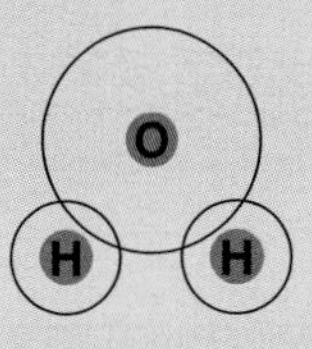

c) ammonia (NH_3)

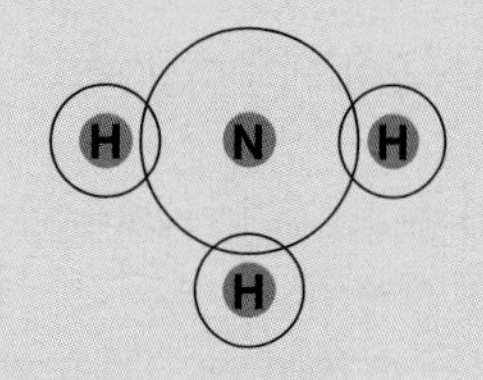

d) methane (CH_4)

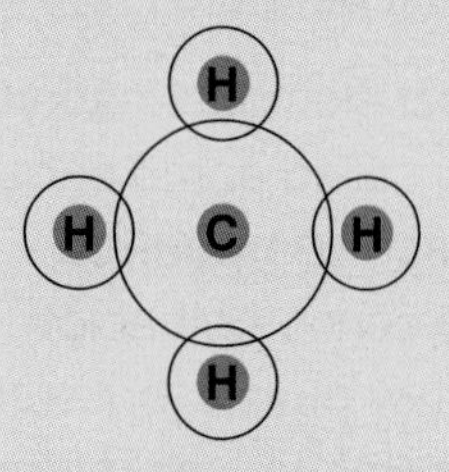

e) Show each of the molecules in a) to d) using 'ball and stick' diagrams.

f) Show each of the molecules in a) to d) using 'dot and cross' diagrams.

g) Will the molecules in a) to d) have high or low boiling points?

5 Look at substances A to D in this table:

	Melting point (°C)	Does it conduct electricity	
		when solid?	when molten?
A	−56	no	no
B	878	no	yes
C	1076	yes	yes
D	2050	no	no

a) Which substance (A, B, C or D) is made up of positive and negative ions?

b) Which is a metal?

c) Which is made up of small molecules?

d) Which has a giant covalent structure?

CHAPTER 16 THE PERIODIC TABLE

16a History of the Periodic Table

Do you ever get the urge to tidy your bedroom;
to put a bit of order in your life?
No? Well, scientists about 200 years ago wanted to put some order into the science of chemistry.
They were looking for the patterns that would make sense of the world. Many believed that God created the world and that's why the patterns must be there.

But at that time, lots of chemical elements had yet to be discovered. Scientists also called some substances elements when they were, in fact, compounds.
So just imagine trying to make sense of that lot!
It's a bit like trying to do a jig-saw puzzle, but without the picture to work from, with some pieces missing and with other pieces that didn't even belong in the box!

How would you like to do a jigsaw with no picture to work from? Some pieces are missing and others don't belong in this jigsaw!
This was the state of Chemistry at the start of the 1800s.

a) What is the difference between an element and a compound?
b) Why do you think scientists 200 years ago mistook some compounds as elements?

Calcium, strontium and barium formed a 'triad'.

At first it was noticed that there were groups of 3 elements that were similar. What was really interesting was that the mass of the middle element in the group of 3 was the average of the other 2 elements. They called these groups 'triads'.
But the pattern could only be found for a few groups of elements.
The rest just didn't follow this pattern, so the search went on.

c) Why were the first 'mini-groups' called triads?

Then around 1865 an English chemist called **John Newlands** announced his discovery. He arranged the known elements in order of their atomic mass and found that every eighth element was similar. But unfortunately his pattern only worked for about the first 15 elements. After that it broke down.
Other scientists at the time mocked his idea.
They thought it was just a coincidence. One said that he'd have better luck arranging the elements in alphabetical order.

John Newlands (1837–1898).

The real breakthrough came a few years later in 1869. It was made by a Russian chemist called **Dmitri Mendeleev**. Like Newlands, he arranged the elements in order of their atomic mass. But he had the same problems.

Dmitri enjoyed playing 'patience' and had made cards for each element. His idea was that the elements would build up row upon row as in his favourite card game.

He struggled with the problem non-stop for days until he fell asleep, exhausted. It was then that he had a dream that revealed to him the final pattern.
Where the pattern broke down, he just left gaps or changed the order so that similar elements did line up.
He called his table the **Periodic Table**.
(The word periodic means 'repeated regularly'.)

But his fellow scientists were hard to persuade. They didn't like the gaps and the need to change the order of masses occasionally to make the table work.

To explain the gaps in the table, Dmitri said that the elements to fit in those spaces hadn't been discovered yet. He predicted the properties of these undiscovered elements using his table.
In 1886 the element germanium was discovered. It closely matched Dmitri's predictions, and the world of science finally accepted his Periodic Table.

Dmitri Mendeleev (1834–1907).
(His name is pronounced 'Dimeetree Mendel-ay-ev'.)
Dmitri was the youngest of 17 children.

The Periodic Table was a scientific breakthrough,
For chemistry made sense, it was easier too.
A Russian named Dmitri was first to spot the pattern,
But some elements were wrong in the spaces they sat in.
'I know,' thought Dmitri, 'I'll just leave some gaps.'
And a stroke of genius had just come to pass.

Some of the elements were as yet undiscovered,
So he made predictions from the properties of others.
A few years later when germanium was found
Scientists agreed his ideas were sound.

Even now we use the Table on which we never dine,
Still based on that discovery in 1869.

Remind yourself!

1 Copy and complete:

The search for a ...... that could explain the properties of chemical ...... resulted in the discovery of the ...... Table by ...... Mendeleev.

2 a) Why were scientists slow to accept Mendeleev's Periodic Table when it was first published?

b) What persuaded them of its usefulness to chemists?

16b The modern Periodic Table

Only one problem remained with Mendeleev's Periodic Table.
Why were a few elements out of order when lined up by their atomic mass?
We must remember that the original table was drawn up well before we knew about what was inside atoms.
Now we know that it is the atomic number that should be used to order the elements.

a) What is the atomic number of an element? (See page 172.)

When we do this, we get the Periodic Table shown below:

Group numbers: 1	2											3	4	5	6	7	0 (or 8)
			H 1 hydrogen														He 2 helium
Li 3 lithium	Be 4 beryllium											B 5 boron	C 6 carbon	N 7 nitrogen	O 8 oxygen	F 9 fluorine	Ne 10 neon
Na 11 sodium	Mg 12 magnesium											Al 13 aluminium	Si 14 silicon	P 15 phosphorus	S 16 sulphur	Cl 17 chlorine	Ar 18 argon
K 19 potassium	Ca 20 calcium	Sc 21 scandium	Ti 22 titanium	V 23 vanadium	Cr 24 chromium	Mn 25 manganese	Fe 26 iron	Co 27 cobalt	Ni 28 nickel	Cu 29 copper	Zn 30 zinc	Ga 31 gallium	Ge 32 germanium	As 33 arsenic	Se 34 selenium	Br 35 bromine	Kr 36 krypton
Rb 37 rubidium	Sr 38 strontium	Y 39 yttrium	Zr 40 zirconium	Nb 41 niobium	Mo 42 molybdenum	Tc 43 technetium	Ru 44 ruthenium	Rh 45 rhodium	Pd 46 palladium	Ag 47 silver	Cd 48 cadmium	In 49 indium	Sn 50 tin	Sb 51 antimony	Te 52 tellurium	I 53 iodine	Xe 54 xenon
Cs 55 caesium	Ba 56 barium	La 57 lanthanum	Hf 72 hafnium	Ta 73 tantalum	W 74 tungsten	Re 75 rhenium	Os 76 osmium	Ir 77 iridium	Pt 78 platinum	Au 79 gold	Hg 80 mercury	Tl 81 thallium	Pb 82 lead	Bi 83 bismuth	Po 84 polonium	At 85 astatine	Rn 86 radon

the alkali metals (Group 1) · the alkaline earth metals (Group 2) · the transition metals · the halogens (Group 7) · the noble gases (Group 0)

Each colour shows a chemical family of similar elements.

You 'read' the Periodic Table like a book:
You start at the top left hand corner, read left to right, going down row after row.

Similar elements line up in columns called **groups**.
They are like 'families' of elements.
They are given numbers, and some have names too.

The rows across the table are called **periods**.
They are numbered from the top, with H and He in the 1st period.
So Li is in Group 1 in the 2nd period.

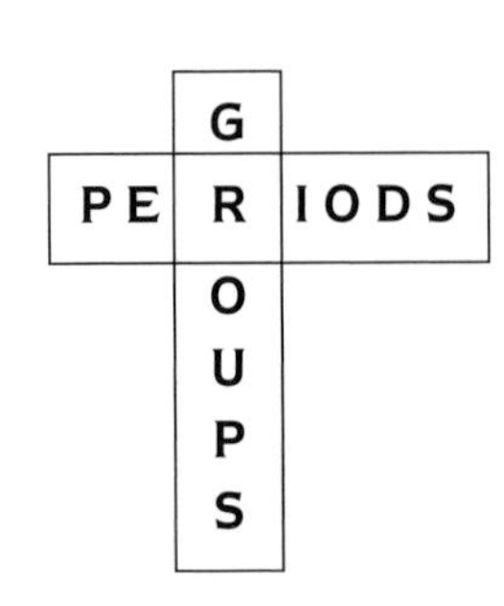

Groups go down, periods go across.

b) What group and period is Al in?

Group 1: The alkali metals

We have met the metals in Group 1 in Chapter 2.

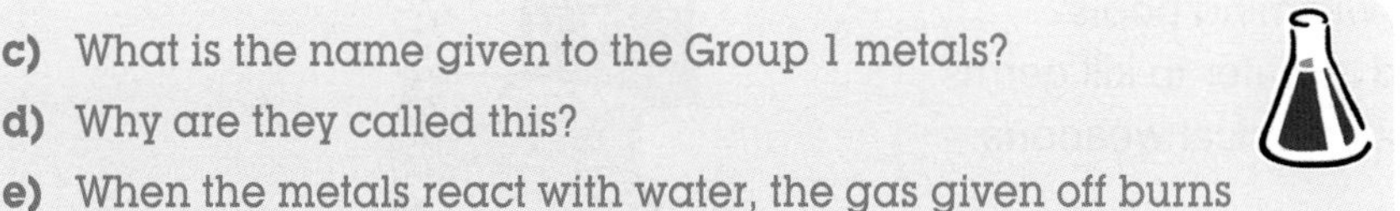

c) What is the name given to the Group 1 metals?

d) Why are they called this?

e) When the metals react with water, the gas given off burns with a 'pop' if you apply a lighted splint. What is the gas?

Look at the atoms of the first 3 elements in Group 1:

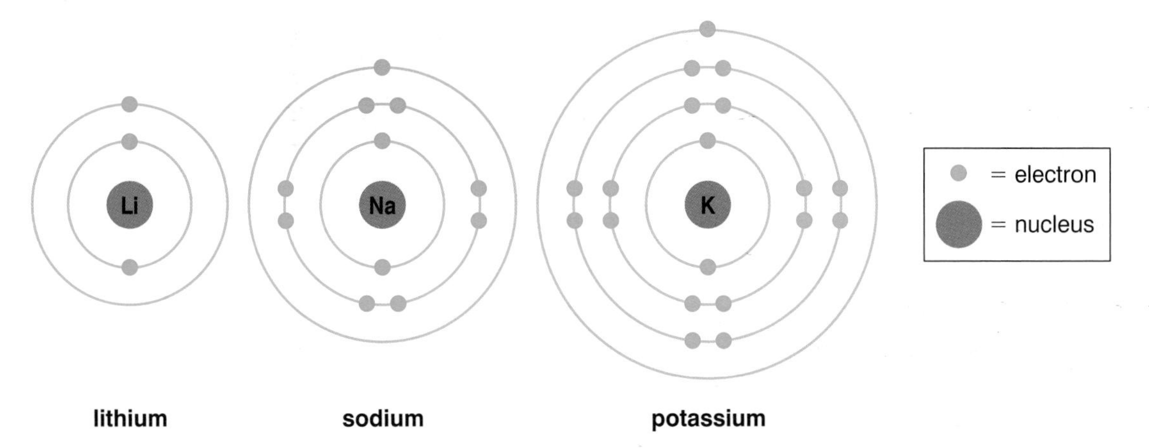

f) What do you notice about the number of electron shells in each atom above?

g) What do you notice about the number of electrons in the outer shell of each atom?

As we go down the Periodic Table each row (period) starts to fill a new shell of electrons.

The number of electrons in the outer shell of an atom tells us the number of the group that element is in.

We often see ***patterns going down a group***. For example, the metals in Group 1 get more reactive as we go down the group. Their melting points get lower as we go down the group.

Remind yourself!

1 Copy and complete:

The elements in the ...... Table are arranged in order of ...... number (not their ......).

The groups are the v...... columns, and the ...... are the horizontal ......

2 a) Explain why the charge on a Group 1 ion is 1+.

b) Find out the melting points of the Group 1 metals and record them in a table.

c) Draw a graph that will show the pattern in their melting points.

16c Group 7: The halogens

Do you know the smell of chlorine gas from swimming pools?
Tiny amounts of its compounds are dissolved in water to kill germs.
It is a highly toxic gas and was one of the first chemical weapons used in World War I.
Chlorine is a halogen from Group 7 in the Periodic Table.
Its formula is Cl_2. It is a **diatomic** molecule.

a) How many chlorine atoms are in a chlorine molecule?

Look at the other members of the halogen 'family' below:
They are written in the order they appear in the Periodic Table.

Halogen (symbol)	Molecule	Colour	State (at 25°C)
fluorine (F)	F_2	pale yellow	gas
chlorine (Cl)	Cl_2	yellow/green	gas
bromine (Br)	Br_2	dark orange/brown (with brown vapour)	liquid
iodine (I)	I_2	grey/black (with violet vapour)	solid

The halogens form 'diatomic' molecules. (They are atomic twins!)

b) Do the halogens get lighter or darker in colour as you go down the group?
c) Look at their states at 25°C:
Do the melting points and boiling points of the halogens get higher or lower as you go down the table?
d) What can you say about all the halogen molecules?

Reactions with metals

The halogens are typical non-metals.
They react with metals to form **salts**.
These are **ionic** compounds in which the halogen forms a halide ion with a **1–** charge.

Halogen	forms the halide ion	example of ionic compound
fluorine	fluoride (F^-)	sodium fluoride
chlorine	chloride (Cl^-)	lithium chloride
bromine	bromide (Br^-)	zinc bromide
iodine	iodide (I^-)	potassium iodide

e) Give another example of a salt formed by each halogen.

Fluorine is the most reactive of all the non-metals.
It reacts with many metals as soon as the gas comes into contact with it. The other halogens react with metals but usually have to be heated.
We find that:

The halogens get *less reactive* as you go down the group.

Reactions with non-metals

The pattern in reactivity is the same in reactions with non-metals.
For example, fluorine will react most vigorously with hydrogen.

hydrogen + fluorine → hydrogen fluoride (HF is a covalent molecule)

Iodine reacts with hydrogen in a slow reversible reaction.
The hydrogen halides formed are molecules that dissolve in water to form acids.

Displacement reactions

We can put the halogens into competition with each other.
The halogens are more stable when they are halide ions.
So if you have a halide ion in solution and another halogen, the less reactive halogen will be **displaced** (or 'kicked out') of the solution. For example,

chlorine + potassium bromide solution → potassium chloride solution + bromine

Chlorine is more reactive than bromine.
So chlorine can displace the bromide ion from solution.
On the other hand:

bromine + potassium chloride solution ↛ no reaction

f) Write the word equation for chlorine added to potassium iodide solution.

Remind yourself!

1 Copy and complete:

The Group 7 elements are called the ……

They are typical …… They get …… reactive as you go down the ……

They form compounds called …… in which their ions have a …… charge.

A more …… halogen can …… a less reactive …… from a …… of its salt.

2 a) Explain why the charge on a Group 7 ion is 1−.

b) Complete this word equation:

bromine + potassium iodide solution →

3 Do some research and make a poster to show the uses of the halogens.

16d Group 0: The noble gases

Have you ever had one of those silver balloons wishing you 'Happy Birthday'? If you have, it was probably filled with **helium** gas. (This is the gas that makes your voice go squeaky!) It is also used to fill airships (or blimps) that you sometimes see advertising things at big events.

Helium is much less dense than air so is used to give airships their lift.

Helium is the first member of Group 0 called the **noble gases**. Look at the whole group of noble gases below:

He – helium
Ne – neon
Ar – argon
Kr – krypton
Xe – xenon
Rn – radon

The noble gases prefer to be alone.

As a group, the noble gases are really boring!
They hardly have any reactions at all. They are said to be **inert**.
The don't even react to form molecules like the halogens.
The noble gases exist as single atoms.
They are called **monatomic** gases.
Look at the noble gas atoms below:

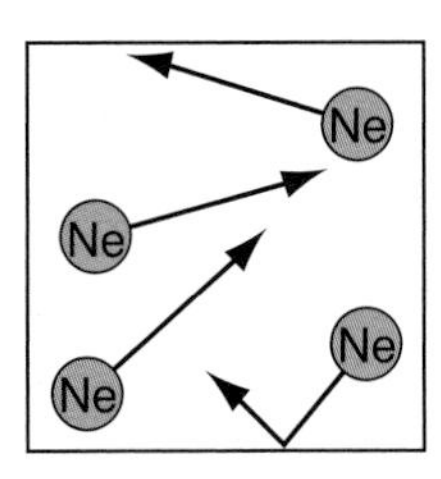

The noble gases are 'monatomic'.

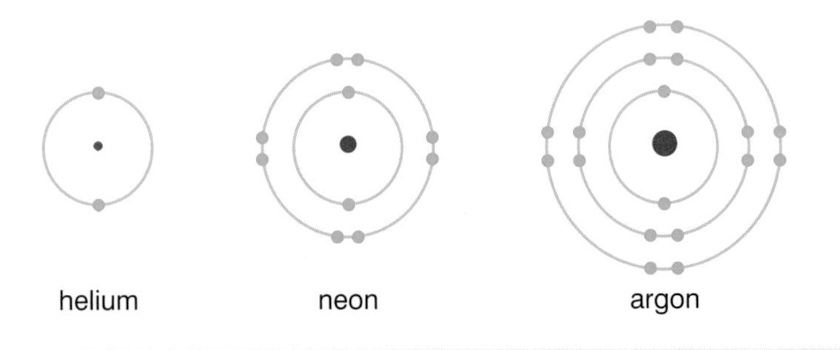

a) What do you notice about the outer shell in each noble gas?

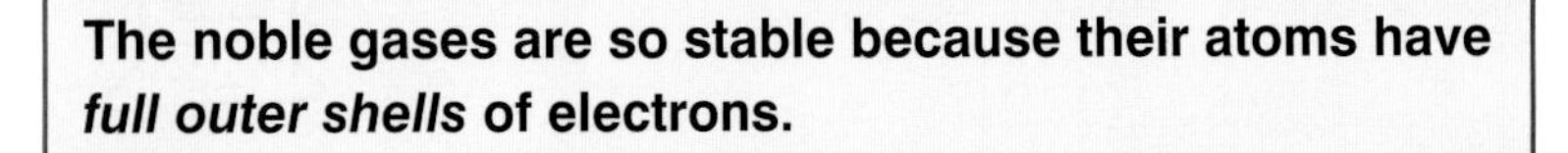
The noble gases are so stable because their atoms have *full outer shells* of electrons.

But their lack of reactivity also makes them useful.
For example, **argon** gas is used inside light bulbs.
If the bulbs were filled with air, the metal filament would react with any oxygen present. It would soon snap.

Argon gas is also used in welding.
It is used around the area where the weld takes place.
It stops oxygen reacting with very hot metals.

b) Why is helium used in airships?
c) Why is argon used in light bulbs?
d) How does argon help in welding metals?

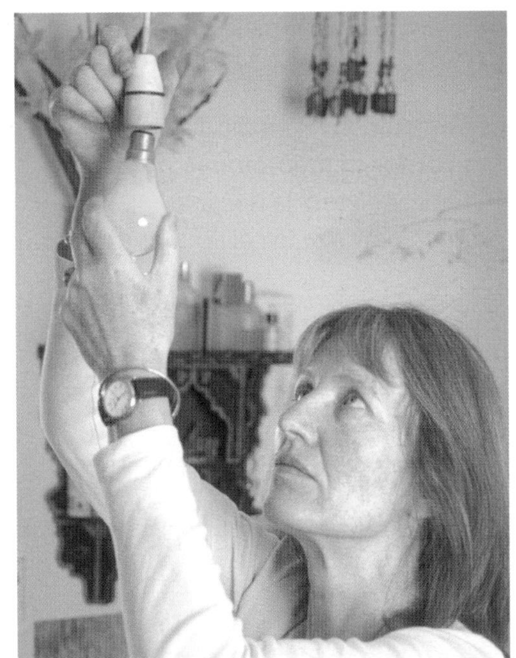

Argon is used in light bulbs.

The noble gases also glow when high voltages are applied to tubes of the gases at low pressure. For example, neon glows bright red. You have probably seen their bright colours in advertising signs:

Krypton gas is used in lasers for eye surgery.
Look at the photo opposite:

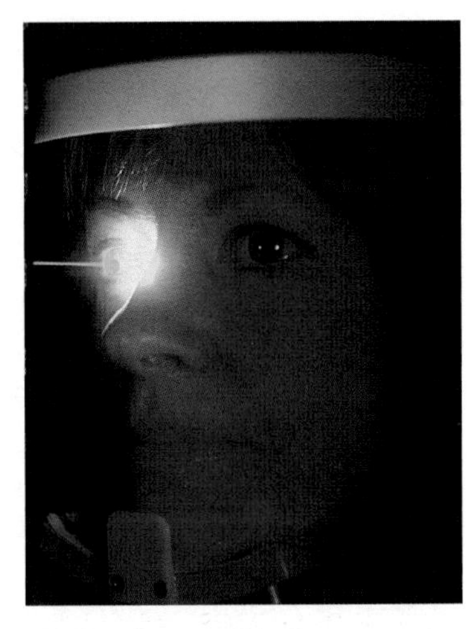

Krypton is used in lasers.

There are some patterns in the noble gases:
- they get more dense as you go down the group,
- their boiling points also increase going down the group.

Remind yourself!

1 Copy and complete:

Group 0 elements are called the ...... gases.
They are very un...... because their atoms have ...... outer shells.

Their boiling points get ...... as you go down the ......, and so does the ...... of the gases.

2 a) Give one use of:

i) helium ii) neon iii) argon iv) krypton.

b) Find out:
i) why some people are worried about radon gas in their houses
ii) which areas of the country are affected by radon gas.

Summary

The modern Periodic Table arranges the elements in ***order of their atomic numbers***.
Similar elements are lined up in columns called **groups**.
Each new row is called a **period**, starting with H and He in the 1st period.

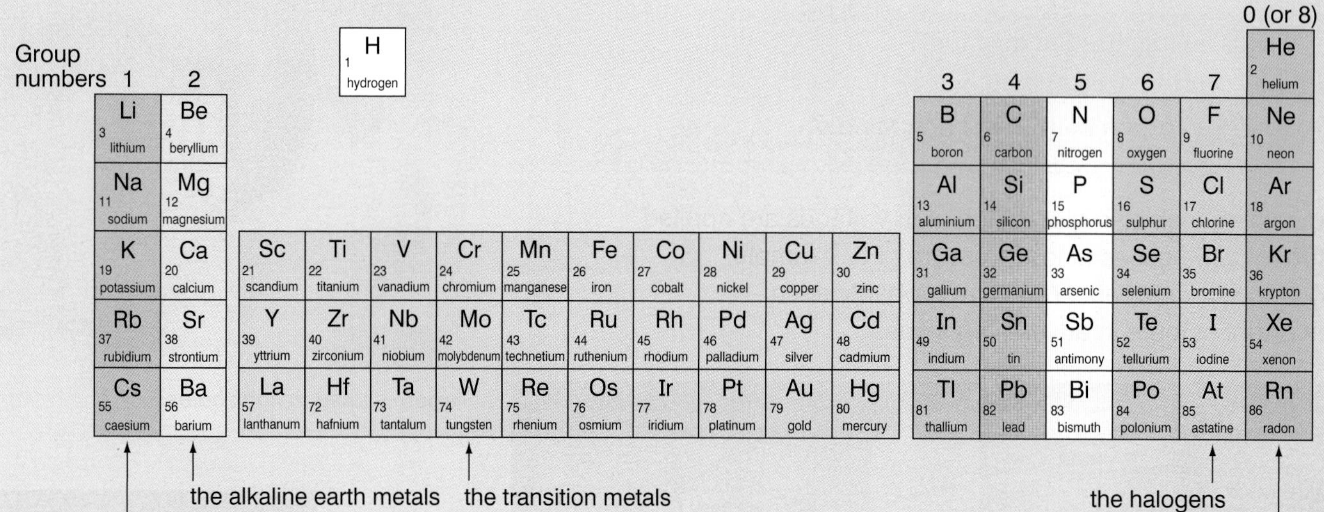

Each colour shows a chemical family of similar elements.

Group 1 elements are called the **alkali metals**.
They react with non-metals to produce ionic compounds, forming ions with a 1+ charge.
They react with water forming a solution of their hydroxide and giving off hydrogen gas.
They get **more reactive** as we go down the group.
Their melting points decrease going down the group.

Group 7 elements are the non-metals called the **halogens**.
They react with metals to produce ionic salts, forming ions with a 1– charge.
They react with other non-metals to form molecules, in which the atoms are joined by covalent bonds.
They all have coloured vapours and exist as diatomic molecules, for example, F_2.
They get **less reactive** as we go down the group.
Their melting points and boiling points increase down the group.

Groups are families of elements. The members of the family are similar but not exactly the same.

Group 0 (or 8) are the very unreactive **noble gases**.
They exist as single atoms (monatomic gases).
Helium is less dense than air and is used in airships and balloons.
The gases are also used inside electric discharge tubes where they glow brightly. For example, neon glows red.
The boiling point and density of a noble gas increase going down the group.

Questions

1 Copy and complete:

The …… Table lines elements up in order of atomic ……

Group …… are the alkali metals. They get …… reactive as we go down the group.

Group 7 are called the …… They get …… reactive as we go down the group.

Group 0 are called the …… gases and they are very un……

2 Compare the properties of Group 1 and Group 7 elements in order to show some of the differences between metals and non-metals.

3 Write down as many patterns as you can going down the following groups:

a) Group 1

b) Group 7

c) Group 0

d) What do all the atoms of the elements in each group have in common?

(Hint: Think about their outer shells.)

4 Look back to Chapter 2 to help you answer this question.

a) Describe what you would see when sodium reacts with water.

b) If you add universal indicator to the solution left after the reaction in a), what pH would you expect?

c) Write a word equation for the reaction of sodium with water.

d) You can collect the gas given off when lithium reacts with water. How could you show that the gas was hydrogen?

e) Which would have the most vigorous reaction with water: lithium, sodium or potassium?

f) Why don't we use rubidium or caesium in schools?

5 Which element from Groups 1, 7 or 0 are described below:

a) It glows bright red in an electric discharge tube.

b) It fizzes around on the surface of water; the gas given off ignites and burns with a lilac flame.

c) It is used in light bulbs.

d) It forms a 1+ ion in table salt.

e) It is used to kill germs in water.

f) It is the least reactive metal in its group.

g) It is made of dark grey crystals and forms a violet vapour when warmed.

h) It is the most reactive non-metallic element.

i) It is used to fill balloons and airships.

j) It is a liquid at 20°C.

6 Look at the boiling points of the halogens below:

Halogen	Boiling point (°C)
fluorine	−188
chlorine	−35
bromine	59
iodine	184

a) Draw a bar chart to show the data above. You might want to draw your base line at −200°C.

b) What pattern do you observe in your graph going down Group 7?

7 a) Imagine you are a scientist writing a letter to Dmitri Mendeleev in 1870. You have seen his Periodic Table but don't agree with his ideas. Include the reasons why you might question his Table.

b) Write another letter to Dmitri in 1886, after the element germanium was discovered. Explain why you now think his Periodic Table is useful after all.

CHAPTER 17

Halides

17a Sodium chloride

We have met **sodium chloride** before in Chapters 5 and 15.
It is the salt we add to food for flavour (or to preserve some foods).
Do you remember 'Nackle'? The formula of sodium chloride is **NaCl**.
We have learned that it is very different from sodium or chlorine.
(See page 178.)

Sodium chloride is the main salt dissolved in the sea.
No wonder we call it **'common salt'**.
It is also found in underground seams as rock salt.
We believe that these deposits were formed millions of years ago when ancient seas dried up.

We can get the salt in three ways:

Evaporating sea water

In hot countries, sea-water is let into shallow lagoons.
Look at the photo opposite:

The heat from the Sun evaporates the water.
The solids that were dissolved in it are left behind.

a) Why is this method not used much in Britain?
b) Would you get pure sodium chloride?
Explain your answer.

Collecting salt from sea-water.

Underground mining

This is a bit like mining for coal.
You dig a shaft down to the seam of rock salt.
Then you dig the salt out, working along the seam.
Large pillars of rock salt are left to support the roof of the mine.
The seam found under Cheshire is up to 2000 metres thick in places!

c) Is rock salt mined from the ground pure sodium chloride?
How can we purify it?

Mining rock salt.

The rock salt dug out can be used to grit the roads in winter. Salt lowers the freezing point of water, making it less likely that ice will form. A lower temperature is needed for ice to form if the roads have been gritted.

Solution mining

This is another way to get salt to the surface from an underground seam of rock salt. This method dissolves the salt in hot water piped down to the seam. The salt solution (called **brine**) is then forced up to the surface.
Look at the diagram below:

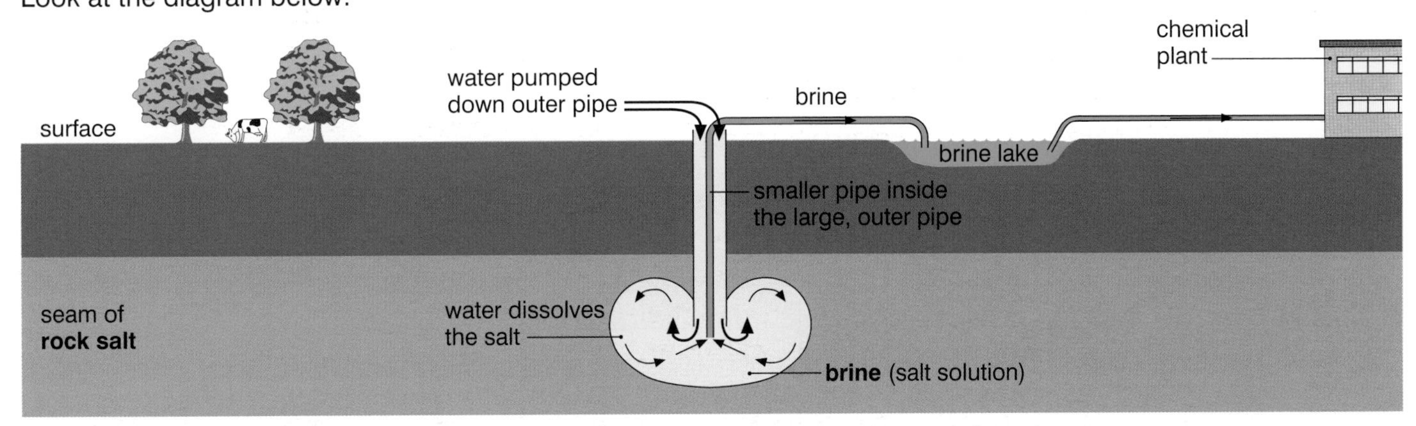

d) How could you get solid sodium chloride from solution mining?

Remind yourself!

1 Copy and complete:

Sodium ...... is also known as ...... salt.

Its chemical formula is ......

We get salt by ...... sea-water, ... it out from underground seams or dissolving it and then pumping the b...... to the surface.

2 Which of these statements are true about sodium chloride?

a) It is a brown solid.

b) It has covalent bonds between atoms.

c) It is made up of white crystals.

d) It has ionic bonds holding ions in a giant lattice.

3 Find out the problems that solution mining can cause in a community.

▸▸▸ 17b Products from brine

Have you eaten any margarine yet today?
Did you know that hydrogen gas was used to make it?
And that the hydrogen can be made from brine?
Margarines are made from oils, such as sunflower oil.
The hydrogen is used to make the oils thicker
so that it spreads well on bread.

a) What is brine?

b) Why is hydrogen used in the manufacture of margarine?

Hydrogen is used in the manufacture of margarine.

You saw on the previous page how we get brine
from underground salt seams.
The brine (salt solution) is stored in a reservoir until it is needed
at the nearby chemical factory. In the factory the brine
is broken down by electricity.

c) What do we call the process whereby a liquid is broken down by electricity? (See page 40.)

You can do the experiment below in the lab:

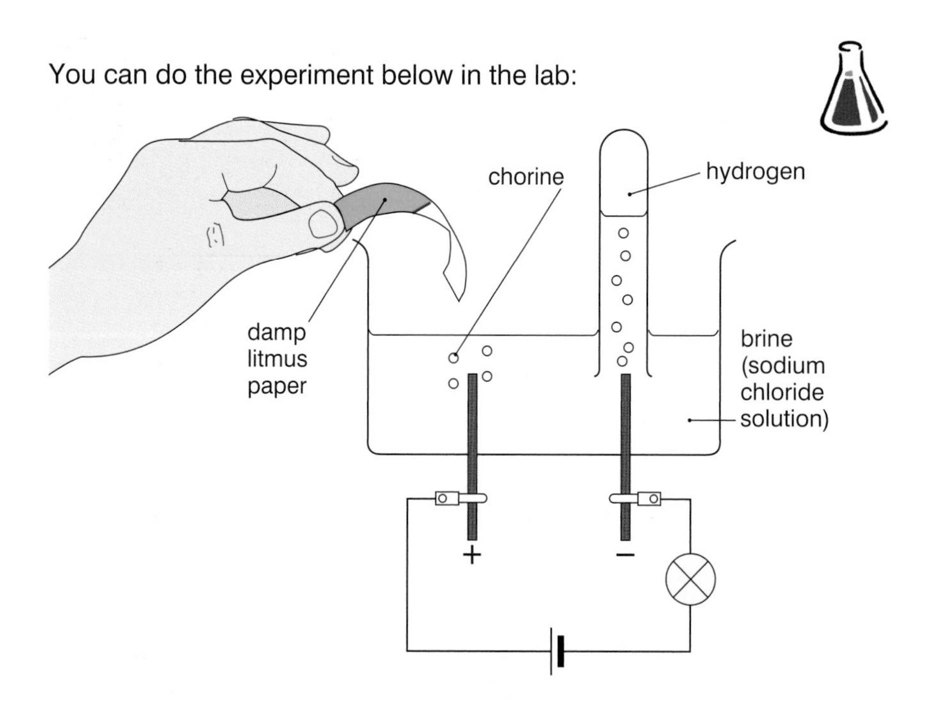

At the negative electrode: hydrogen gas is given off.
(This gas will burn with a squeaky 'pop' when a lighted splint is held in it.)

At the positive electrode: chlorine gas is given off.
(We can test for this gas as it bleaches damp litmus paper.)

The solution around the negative electrode turns ***alkaline***.
(You can test this with Universal Indicator solution.)
This is because **sodium hydroxide solution** is formed.

d) What colour will the Universal indicator turn?

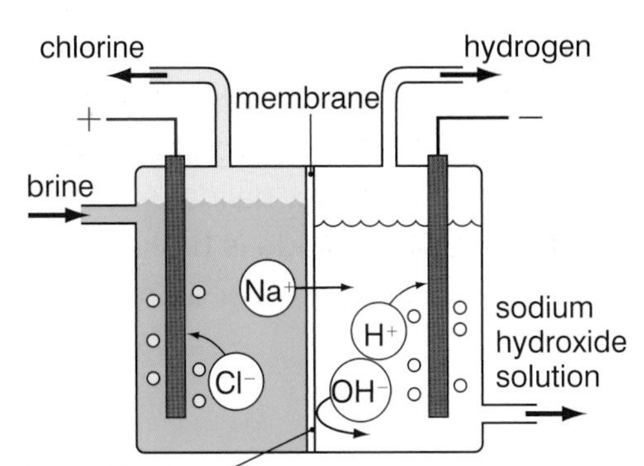

In industry the electrolysis takes place in many cells, all working at once. The most modern cells are called membrane cells. Look at the diagram opposite:
A plastic barrier (membrane) is used to:
– keep the hydrogen and chlorine gas separate
– stop the chlorine reacting with the sodium hydroxide
– only let positive ions through, so sodium hydroxide can be tapped off from the cell.

Look at the products we can make using chlorine, hydrogen and sodium hydroxide:

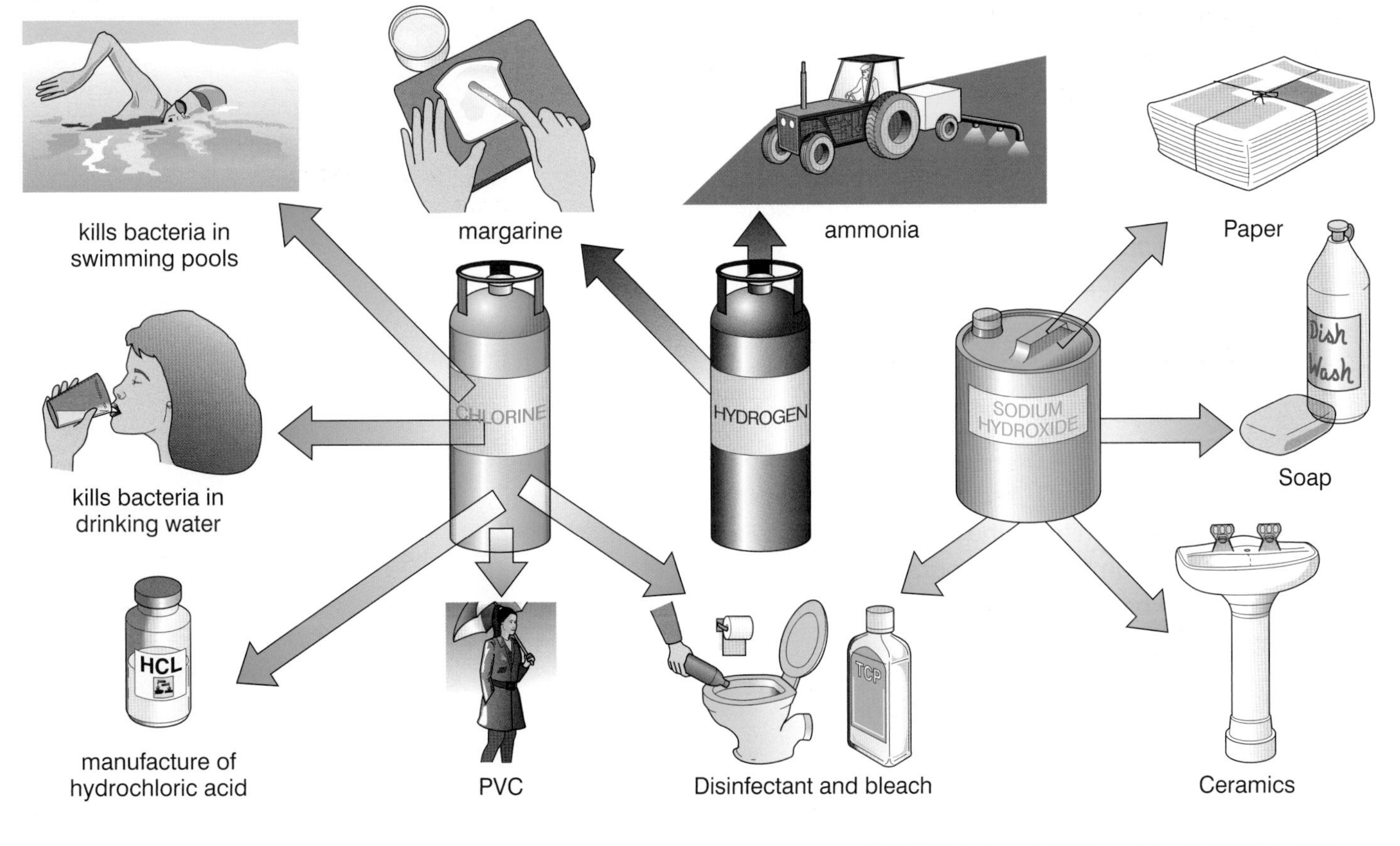

Remind yourself!

1 Copy and complete:

When brine is electrolysed, we get hydrogen, ... and sodium ...... solution formed. We can test for the gases given off with a ...... splint that ...... (at the ...... electrode), and with damp indicator paper that is ...... (at the ...... electrode).

2 a) Make a list of the things you use each day that have been made from the chemicals we get from brine.

b) Why is the chemical industry in Cheshire known as the chlor-alkali industry?

17c Silver halides

Photographic film contains silver halides.

As you know from the last chapter, halides are compounds of the halogens.

Whenever you look at a photograph, you can thank the silver halides. Photographic film or paper contains a mixture of the silver halides. When they are exposed to light the silver halide is broken down. Silver ions are reduced to form silver atoms. These are deposited on the film.

For example:

$$\text{silver bromide} \xrightarrow{\text{light}} \text{silver} + \text{bromine}$$

a) Write the equation for silver chloride breaking down.

Silver bromide was used originally on 'black and white' film, but in colour film the mixture is more complex.

The same reaction happens with X-rays and radioactivity. In fact, in 1896 radioactivity was discovered thanks to silver bromide on a photographic plate.

Henri Becquerel left a rock on a photographic plate inside a dark drawer. When he came to use the plate he noticed it had 'fogged up'. This was caused by radioactive atoms in the rock.

Have you ever had an X-ray? The X-rays do not pass through bones so they appear white on the film. The areas where X-rays do pass through hit the photographic film. The silver bromide breaks down and the dark patches are caused by the silver metal deposited.

b) Why are X-rays useful for doctors?

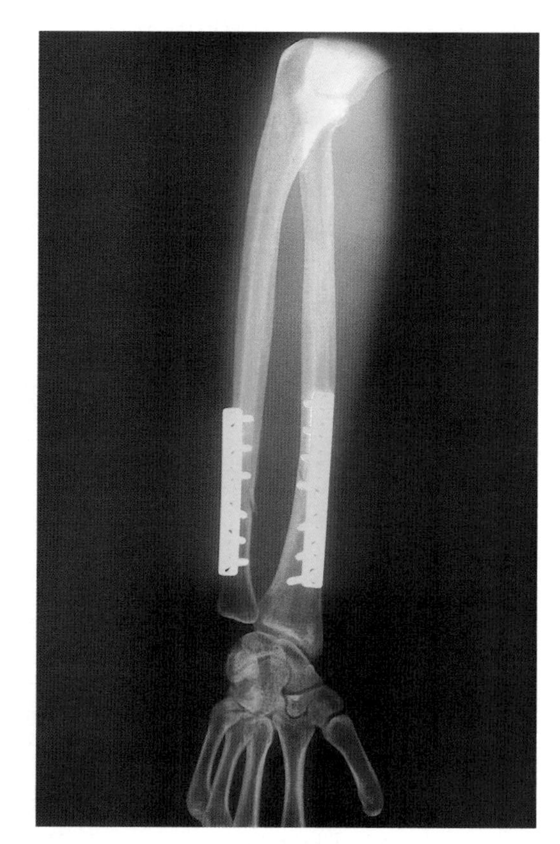
What does this X-ray plate show?

Remind yourself!

1 Copy and complete:

The silver ...... are used in photographic ...... and paper. They are broken down in ...... to form ...... and the halogen. The silver ions are r...... to silver atoms.

2 a) Besides light, what else causes silver halides to break down?

b) Find out about the early work on radioactivity and the scientists who investigated it.

Summary

Sodium chloride (common salt) is found in the sea and underground as rock salt.
It can be pumped up from underground as brine (salt solution).

The brine is electrolysed in industry to form hydrogen, chlorine and sodium hydroxide solution.

The hydrogen is given off from the negative electrode.
(Hydrogen makes a 'pop' – a squeaky explosion – with a lighted splint.)
Chlorine gas is given off from the positive electrode.
(Chlorine bleaches damp litmus paper.)

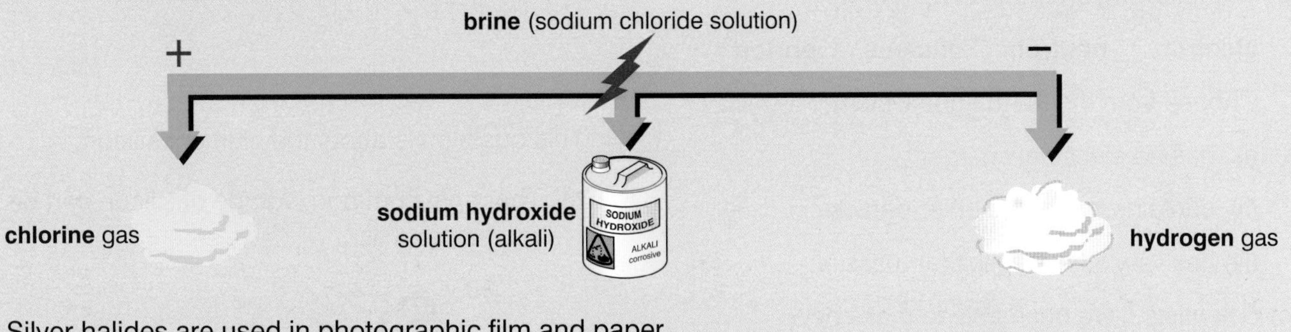

Silver halides are used in photographic film and paper.
They are broken down by light, X-rays or radioactivity.

Questions

1 Copy and complete:

Sodium chloride solution is known as ......

This is an important raw ...... in the chemical industry, producing ......, chlorine and ...... hydroxide solution.

The silver ...... are used in p...... They are broken down by ......, X-rays and ......

2 a) Write down the equation for the reaction that happens if silver iodide is left in a test tube on the window sill.

b) What difference would you expect if the silver iodide was kept in a dark cupboard?

c) How did silver bromide help in the discovery of radioactivity?

3 In the electrolysis of brine:

a) Which gas is given off at the negative electrode?
How could you test for this gas?

b) Which gas is given off at the positive electrode?
How could you test for this gas?

c) Which electrode does sodium hydroxide solution form around?
What will be the pH of sodium hydroxide solution?

d) Draw spider diagrams to show the uses of chlorine, hydrogen and sodium hydroxide.

e) The membrane cell is used in industry.
Find out about another type of cell used.

Further questions on Chemical patterns and bonding

Inside atoms

1 This question is about elements and atoms.

(a) About how many different elements are found on Earth?
Choose the correct number from the list below:

40 50 60 70 80 90

(1)

(b) The following are parts of an atom:

electron neutron nucleus proton

Choose from the list the one which:

(i) has no electrical charge;

(ii) contains two of the other particles;

(iii) has very little (negligible) mass. (3)

(c) Scientists have been able to make new elements in nuclear reactors. One of these new elements is fermium. An atom of fermium is represented by the symbol below.

$$^{257}_{100}\mathrm{Fm}$$

(i) How many protons does this atom contain?

(ii) How many neutrons does this atom contain? (2)

(AQA 1999)

2 The diagram shows atoms of carbon-12 and carbon-14.
The symbols ○ ● X represent the particles found in atoms.

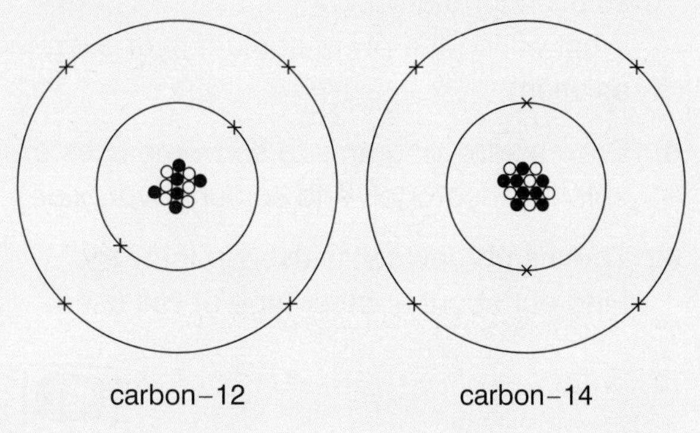

Use words from the list to complete the sentences.

electrons isotope molecule neutrons

nucleus orbit protons

Particles with the symbol **X** in the diagram are called ......

Particles with the symbol ○ and ● together make up the ...... of an atom.

Carbon-14 has more ...... than carbon-12.

(3)

(AQA 2001)

3 This question is about the element silicon.

(a) The most common isotope of silicon can be represented as:

$$^{28}_{14}\mathrm{Si}$$

(i) How many protons and neutrons are there in an atom of this isotope?

Number of protons ....
Number of neutrons ....

(2)

(ii) Silicon has other isotopes. Which word completes the following sentence?

Isotopes of the same element have different numbers of ....

(1)

(b) Copy and complete the diagram below to show the arrangement of all the electrons in an atom of silicon.

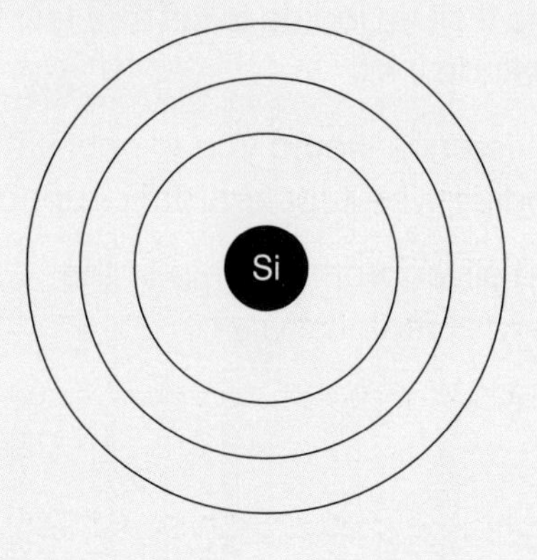

(2)

(AQA 2000)

Bonding

4 This question is about magnesium oxide and how it is formed from atoms of magnesium and oxygen.

(a) Finish the sentences by choosing the **best** words from this list.

atoms

covalent

giant

high

ionic

ions

low

molecular

Magnesium oxide is an example of a substance with i) …… bonding.

It has a ii) …… structure made up of iii) …….

Strong forces between the particles in the structure cause magnesium oxide to have a iv) …… melting point. (4)

(b) The table gives information about the electron arrangement in a magnesium atom and an oxygen atom.

Atom	Electron arrangement
Mg	2,8,2
O	2,6

Describe the changes in electron arrangement that take place when magnesium oxide is formed from magnesium and oxygen. (4)

(OCR Nuffield 1999)

5 **(a)** Use the Periodic Table on page 192 to complete parts (i), (ii), (iii) and (iv).

(i) The mass number of potassium is ……. (1)

(ii) The number of protons present in an atom of aluminium is ……. (1)

(iii) The number of neutrons present in an atom of fluorine is ……. (1)

(iv) Using **X** to represent an electron, copy and complete the following diagram to show the electronic arrangement for an atom of **sulphur**. (1)

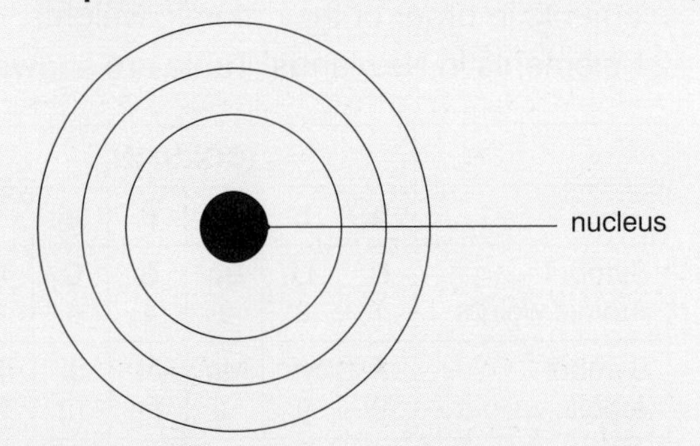

(b) Chlorine atoms have seven electrons in their outermost orbit (shell). State **two** different ways in which chlorine atoms can form a chemical bond. (2)

(WJEC)

6 Aluminium oxide contains aluminium ions (Al^{3+}). The atomic number of aluminium is 13. The mass number of aluminium is 27.

(i) Copy and complete the table to show the number of protons, neutrons and electrons in an aluminium atom (Al) and an aluminium ion (Al^{3+}).

Formula of particle	Number of protons	Number of neutrons	Number of electrons
Al	....	....	....
Al^{3+}	....	....	....

(4)

(ii) Name the type of bonding present in aluminium oxide. (1)

(iii) State ONE physical property you would expect aluminium oxide to have. (1)

(EDEXCEL 2000)

7 This question is about atoms, molecules and ions.

Answer **a)** to **c)** from this list:

Ne **Cu^{2+}** **CO_2** **H_2O** **Cl^-** **Br^-**

(a) Write down the formula of an ion. (1)

(b) Write down the formula of a molecule. (1)

(c) Write down the symbol for a noble gas (1)

(OCR Suffolk 1999)

Further questions on Chemical patterns and bonding

The Periodic Table

8 John Newlands attempted to classify the elements in 1866. He tried to arrange all the known elements in order of their atomic weights. The first 21 elements in Newlands' Table are shown below.

	COLUMN						
	a	b	c	d	e	f	g
Symbol Atomic weight	**H** 1	**Li** 2	**Be** 3	**B** 4	**C** 5	**N** 6	**O** 7
Symbol Atomic weight	**F** 8	**Na** 9	**Mg** 10	**Al** 11	**Si** 12	**P** 13	**S** 14
Symbol Atomic weight	**Cl** 15	**K** 16	**Ca** 17	**Cr** 18	**Ti** 19	**Mn** 20	**Fe** 21

Use the Periodic Table on page 192 to help you answer these questions.

(a) In two of Newlands' columns, the elements match the first three elements in two Groups of the modern Periodic Table.
Which two columns, **a** to **g**, are these? (1)

(b) (i) A Group in the modern Periodic Table is completely missing from Newlands' Table. What is the number of this Group? (1)

(ii) Suggest a reason why this Group of elements is missing from Newlands' Table. (1)

(c) Give **one** difference between iron, Fe, and the other elements in column **g** of Newlands' Table. (1)

(d) Give the name of the block of elements in the modern Periodic Table which contains Cr, Ti, Mn and Fe. (1)

(AQA 2000)

9 Choose from the list the **Group** of the Periodic Table to which each element belongs.
(Page 192 will help you to answer this question.)

Group 1	Group 2	Group 3	Group 4	Group 5	Group 6	Group 7	Group 0

(a) A colourless gas which relights a glowing splint.

It is in Group .... (1)

(b) A green/yellow gas which is a non-metal. It forms ions with a 1− charge.

It is in Group .... (1)

(c) A colourless gas which exists as individual atoms. It is very unreactive.

It is in Group .... (1)

(d) A shiny metal when cut. It reacts violently with cold water. It forms ions with a 1+ charge.

It is in Group .... (1)

(e) A colourless gas which makes up 78% by volume of the air.

It is in Group .... (1)

(f) The atoms of this element have 12 protons in the nucleus.

It is in Group .... (1)

(AQA 1999)

10 **(a)** Use the elements in the box to complete the table.

aluminium	argon	chlorine
hydrogen	lithium	magnesium
oxygen	silicon	sulphur

Write the name of the element that best matches each description. You may use each element once or not at all.

Description	**Name of element**
This element is a low density metal that reacts quickly with water.	(i)
This element is a reactive gas that when added to water makes a bleach.	(ii)
This element is an unreactive gas that is used in light bulbs.	(iii)
This element reacts with copper to form copper oxide.	(iv)
This element has the chemical symbol S.	(v)

(5)

(AQA SEG 1998)

Further questions on Chemical patterns and bonding

11 Lithium is a very reactive metal.

(a) Lithium reacts with cold water.

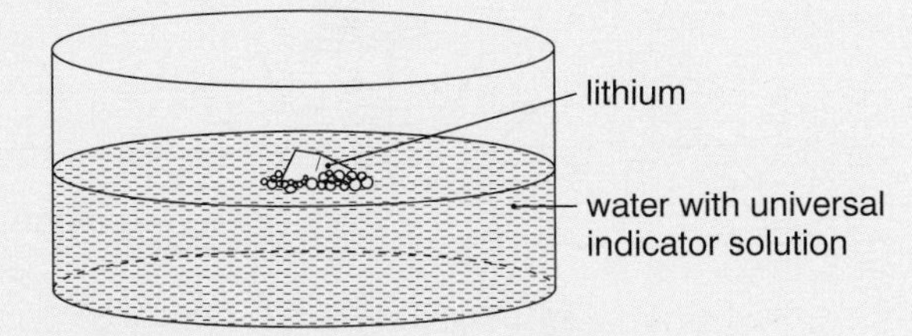

(i) Which **physical** property of lithium is seen during this reaction? (1)

(ii) Which **chemical** property of lithium will be shown by the universal indicator? (1)

(b) Copy and complete the sentence by writing in the missing numbers.

Lithium has an atomic number of 3 and a mass number of 7.

This means that an atom of lithium has …… protons, …… electrons and …… neutrons. (3)

(AQA SEG 2000)

12 The diagram shows some of the elements in Groups 1 and 7 of the Periodic Table.

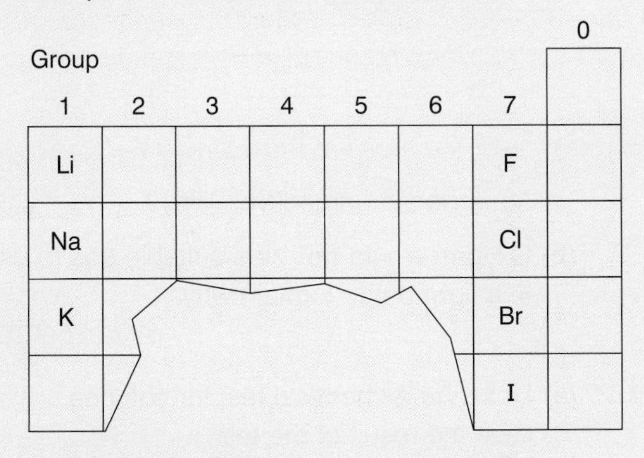

(a) The elements in Group 1 have similar chemical properties.

Describe **one** chemical reaction which shows that lithium, sodium and potassium react in the same sort of way.
You should say what you would react them with and what substances would be produced.

- What you would react them with? (1)
- What substances would be produced? (2)

(b) All the elements in Group 7 react with hydrogen.

Fluorine reacts in the dark, explosively, at very low temperatures.
Chlorine reacts explosively in sunlight, at room temperature.
Bromine, in light, only reacts if heated to about 200°C.

Suggest the conditions needed for hydrogen and iodine to react.
Give reasons for your answer. (2)

(AQA 1999)

13 Chlorine, bromine and iodine belong to Group VII in the Periodic Table.

(a) The table below shows some properties of Group VII elements.
Complete the table by putting in the correct **state** of each element at room temperature (20°C).

Element	*Melting Point (°C)*	*Boiling Point (°C)*	*State (solid, liquid or gas)*
bromine	−7	59	i) …………
chlorine	−101	−35	ii) …………
iodine	114	184	iii) …………

(3)

(b) Give **two physical** properties the **vapours** of Group VII elements have in common. (2)

(c) (i) Explain, in terms of electronic structure, why Group VII elements have similar **chemical** properties. (1)

(ii) How does the reactivity of Group VII elements change in going **down** the group from chlorine to iodine? (1)

(d) Sodium burns vigorously in chlorine to form sodium chloride.
Complete and balance the **symbol** equation for the reaction.

$$\ldots\ldots \text{Na} + \ldots\ldots \rightarrow \ldots\ldots \text{NaCl}$$

(2)

(e) Give **one** large scale use of chlorine. (1)

(WJEC)

14 The first airships were filled with hydrogen gas. Now helium gas is used in airships.

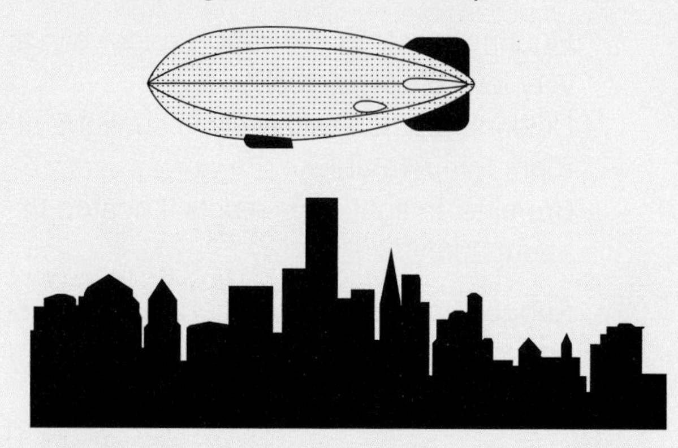

(a) Some chemical symbols are given in the box.

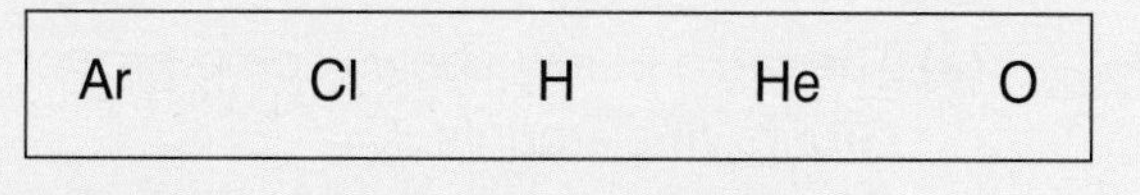

Ar	Cl	H	He	O

Choose the chemical symbol for:

i) hydrogen ii) helium (2)

(b) Write down **one** property that makes hydrogen and helium useful for airships.

Properties of hydrogen and helium
colourless
low density
no smell

(1)

(c) Explain why helium is now used in airships and why hydrogen is no longer used. (2)

(AQA SEG 2000)

15 Choose from these elements to complete the table below.

argon bromine hydrogen iron sodium

(a) an alkali metal		(1)
(b) a gas used in filament lamps (light bulbs)		(1)
(c) an element with a coloured vapour		(1)

(AQA 1999)

16 The diagram shows a light bulb.

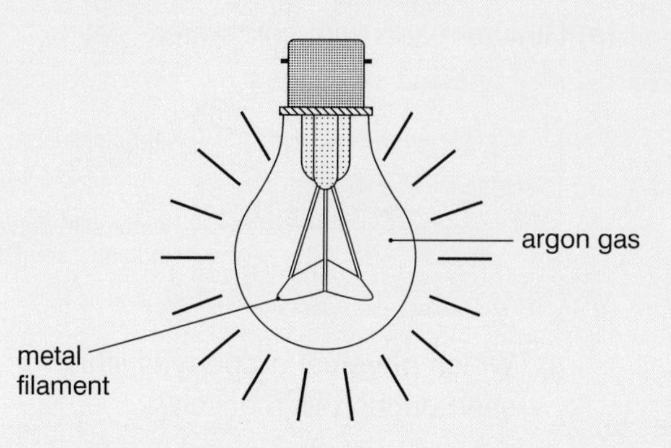

(a) (i) An argon atom has the structure shown below. Use the words in the box to label the particles in the atom. Each word should only be used **once**.

electron	neutron	proton

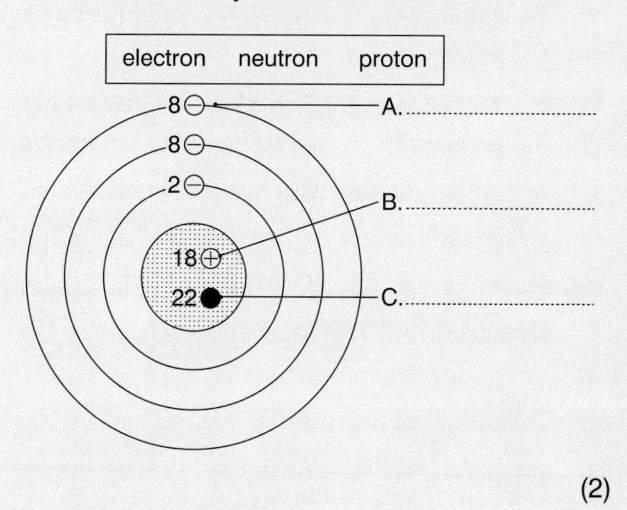

(2)

(ii) Argon is unreactive. Why? (1)

(b) Oxygen would **not** be a suitable gas to use in a light bulb. Explain why. (2)

(AQA SEG 2000)

17 **(a)** Describe a chemical test for chlorine. Give the result of the test. (2)

(b) What happens when chlorine gas is added to potassium iodide solution? You should describe what you would see **and** state what substances are formed. (3)

(AQA 1999)

18 This question is about metals and non-metals.

Use the Periodic Table on page 192 to help you to answer these questions.

(a) Write down the names of **two** elements which are **metals**.

...... and ...... (1)

(b) Write down the names of **two** elements which are **non-metals**.

...... and ...... (1)

(c) The element **selenium** has the symbol **Se**.

Find selenium on the Periodic Table.

What type of element is selenium? (1)

(d) Write about the **differences** in properties between metals and non-metals.

Write about both **physical** and **chemical** properties. (4)

(e) When copper (Cu) is burnt in oxygen (O_2), copper oxide (CuO) is made.

Write a balanced symbol equation for this reaction. (2)

(OCR Suffolk 1999)

19 Sodium (atomic number 11) and potassium (atomic number 19) are both in Group 1 of the Periodic Table because they have the

(a) same number of electrons;

(b) same number of electrons in their outer shells;

(c) same number of neutrons;

(d) same number of protons.

(AQA SEG 2000)

20 Which property **decreases** down Group 7 of the Periodic Table?

(a) Atomic number;

(b) Boiling point;

(c) Melting point;

(d) Reactivity.

(AQA SEG 2000)

Halides

21 The diagram shows electrolysis of sodium chloride solution.

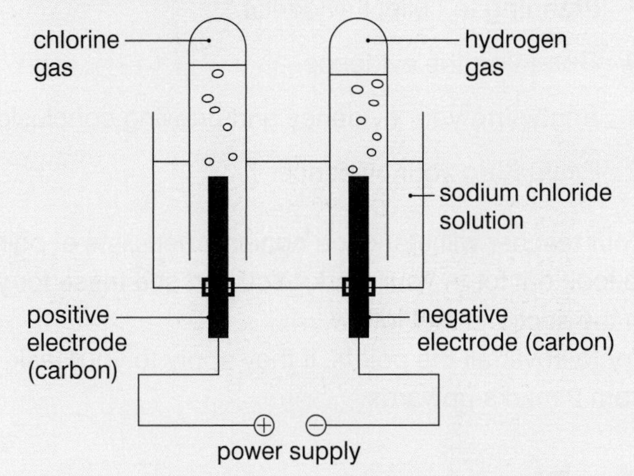

(a) Name the substance left in solution at the end. (1)

(b) Complete the table for the two gases produced during electrolysis.

GAS	TEST FOR THE GAS	SUBSTANCES MANUFACTURED FROM THE GAS
Hydrogen	burns with a squeaky pop	A. .... B. ....
Chlorine	C.	1. bleach 2. PVC

(3)

(AQA 1999)

22 Photographic film can be made by coating paper with silver chloride.

Explain what happens to the silver chloride when the film is exposed to light. (2)

(AQA 2000)

Doing your coursework

GCSE exams have 20% for the marks awarded for coursework.
Your teacher has to assess your practical skills.
You are given marks in 4 areas:

P **Planning** to collect evidence.

O **Obtaining** the evidence.

A **Analysing** your evidence and drawing conclusions.

E **Evaluating** your evidence.

Your teacher will mark you against checklists of points to look out for in your work. You can see these for yourself in the sections that follow.
Try to cover all the points, if they apply to your task, working from 2 marks upwards.

Practical skills are important!

P Planning

Choosing apparatus

It is important to use the most suitable equipment.
For example, if you are measuring 50 cm^3 of water, you should use a measuring cylinder. It is not a good idea to just use a beaker with a 50 cm^3 mark on it. Why not?

Deciding on how many readings

You will need to think about how many measurements or observations to make in your experiments.
If you plan to show your results on a line graph, aim to collect 5 different measurements.
And if the measurements are tricky to make, you should repeat them. Taking the average (mean) of the measurements will make them more **reliable**.

Safety

Make sure your tests are safe.
You must check to see if the chemicals you plan to use or make are hazardous.
The symbols on page 123 will help.

Checklist for skill P PLANNING YOUR WORK	
Candidates:	**Marks awarded**
• plan a simple method to collect evidence	**2**
• plan to collect evidence that will answer your questions • plan to use suitable equipment or other ways to get evidence	**4**
• use scientific knowledge to: – plan and present your method – identify key factors to vary or control – make a prediction if possible • decide on a suitable number and range of readings (or observations) to collect	**6**
• use detailed scientific knowledge to: – plan and present your strategy (the approach you have decided on) – aim for precise and reliable evidence – justify your prediction if you made one • use information from other sources, or from preliminary work in your plan	**8**

O Obtaining your evidence

Making accurate measurements and observations

Accuracy is important, as well as taking care in checking your results.

Common mistakes include:

- not checking your balance is on zero when measuring mass,
- spilling powders before, during or after finding their mass,
- not reading to the ***bottom*** of the meniscus (curve) when measuring volumes of liquid.

You should consider whether using **data-loggers** will improve the quality of the evidence you collect.

If one of your results seems unusual, make sure you repeat it. If you find that it was an error, you don't have to include it in your final results.
(But do comment on it in your evaluation!)

Checklist for skill O OBTAINING YOUR EVIDENCE	
Candidates:	**Marks awarded**
• use simple equipment safely to collect some results	**2**
• make adequate observations or measurements to answer your questions • record the results	**4**
• make observations or measurements, – with sufficient readings, – which are accurate, and – repeat or check them if necessary • record the results clearly and accurately	**6**
• carry out your practical work – with precision and skill, – to obtain and record reliable evidence, – with a good number and range of readings	**8**

Recording your results

You will often record your results in a table.
In the first column of your table you put the thing (variable) that you changed in your experiment.
In the second column you put the thing (variable) that you judged or measured.
Look at the examples below:

Metal	Time to finish reacting (s)
zinc	220
magnesium	55
calcium	30
iron	350

You changed the metal each time. *You measured how long the reaction took.*

Concentration of acid (M)	Volume of gas given off (cm^3)
0.0	0
0.2	25
0.4	47
0.6	77
0.8	95

You changed the concentration of acid each time. *You measured how much gas was given off.*

A Analysing and drawing conclusions

Drawing graphs and bar charts

Once you have recorded your results in a table, drawing a graph will show you any patterns.

Whether you draw a bar chart or a line graph depends on your investigation.
Here's a quick way to decide which to draw:

> If the thing (variable) you change is described in words, then draw a **bar chart**.
> If the thing (variable) you change is measured, then draw a **line graph**.

Let's look at the results from the tables on the previous page:

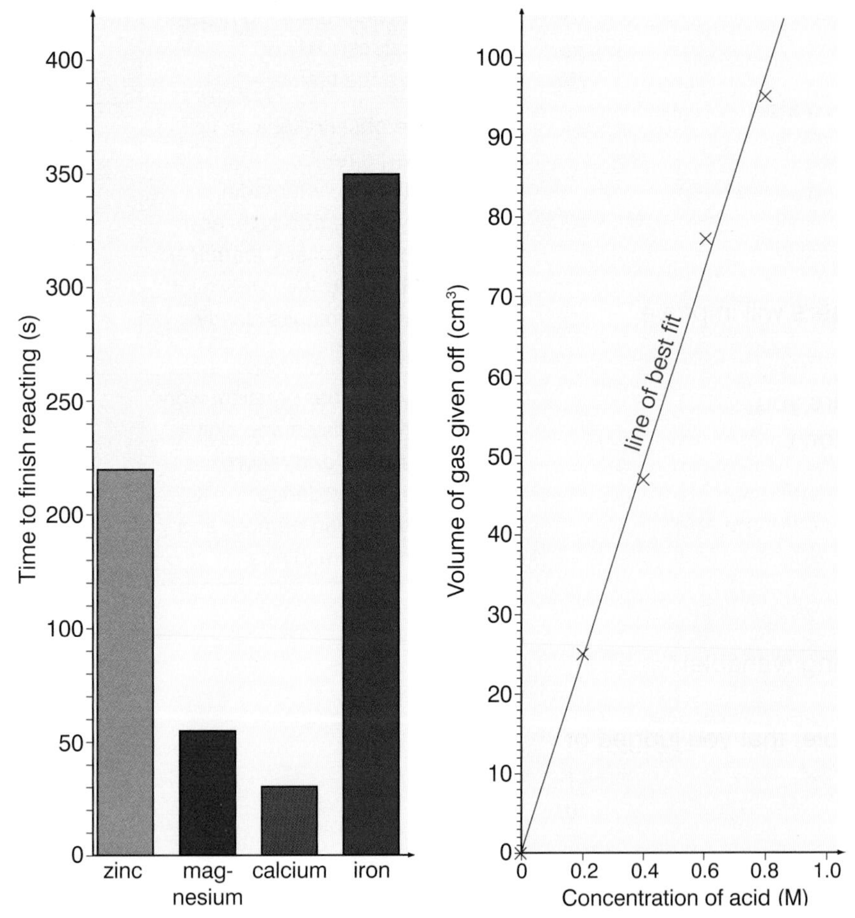

Notice that the thing (variable) you change always goes along the bottom of your graph. The thing (variable) you measure goes up the side.

You can then see if there is a link between the two variables. For example, the higher the concentration of the acid, the more gas was given off. See the line graph above.

Then try to explain any patterns you spot on your graphs using the ideas you have learned about in science.
If you made a prediction, refer back to it in your conclusion.

Checklist for skill A ANALYSING YOUR EVIDENCE	
Candidates:	**Marks awarded**
• state simply what was found out	**2**
• use your results from simple diagrams, charts or graphs to help explain your evidence • spot any trends and patterns in the results	**4**
• draw and use diagrams, charts, graphs (with a line of best fit), or calculate answers from the results • draw a conclusion that fits in with your evidence and explain it using your scientific knowledge	**6**
• use detailed scientific knowledge to explain a valid conclusion (drawn from analysing your evidence) • explain how well the results agree or disagree with any prediction made earlier	**8**

E Evaluating

When you have drawn your conclusions, you should now think about how well you did your investigation.

Ask yourself these questions to see if you could have improved your investigation:

- Were my results accurate?
- Did any seem 'strange' compared to the others? These are called **anomalous** results.
- Should I have repeated some tests to get more reliable results? Could I improve the method I used?
- Did I get a suitable range of results? The range is the spread of values you chose.

  For example: if you were seeing how temperature affected something, choosing to do tests at 20°C, 21°C and 22°C would not be a good range to choose!
- If there is a pattern in my results, is it only true for the range of values that I chose? Would the pattern continue beyond this range?
- Would it be useful to check your graph by taking readings ***between*** points?

  For example: if you have a sudden change between two points, why not do another test to get a point half way between? This would check the shape of the line you get.
- How would I have to change my investigation to get the answers to the questions above?

Checklist for skill E EVALUATING YOUR EVIDENCE	
Candidates:	**Marks awarded**
• make a relevant comment about the method used or the results obtained	**2**
• comment on the accuracy of the results, pointing out any anomalous ones • comment on whether the method was a good one and suggest changes to improve it	**4**
• look at the evidence and: – comment on its reliability, – explain any anomalous results, – explain whether you have enough to support a firm conclusion • describe, in detail, further work that would give more evidence for the conclusion	**6**

Key Skills

As you study Science, you will need to use
some general skills along the way.
These general learning skills are very important,
whatever subjects you take or job you go on to do.

The Government has recognised just how important
the skills are by introducing a new qualification.
It is called the **Key Skills Qualification**.
There are 6 key skills:

- **Communication**
- **Application of number**
- **Information Technology (IT)**
- **Working with others**
- **Problem solving**
- **Improving your own learning**

The first 3 of these key skills will be assessed by exams
and by evidence put together in a portfolio.
You can see what you have to do to get the first level
in the sections below.

Communication

In this key skill you will be expected to:

- **Hold discussions**
- **Give presentations**
- **Read and summarise information**
- **Write documents**

You will do these as you go through your course,
and producing your coursework will help.

Look at the criteria below for communication:

What you must do ...
Take part in discussions.
Read and obtain information.
Write different types of document.

Application of number

In this key skill you will be expected to:

- **Obtain and interpret information**
- **Carry out calculations**
- **Interpret and present the results of calculations**

What you must do ...
Interpret information from different sources.
Carry out calculations.
Interpret results and present your findings.

Information Technology

In this key skill you will be expected to:

- **Use the internet and CD ROMs to collect information**
- **Use IT to produce documents to best effect**

What you must do ...
Find, explore and develop information.
Present information, including text, numbers and images.

Revising and doing your exams

Work out your best way of learning!

When you walk into your Science exam, you will already have your coursework marks completed.
If you do Modular Science, you will also have your test marks.
But your final exam is still the biggest part of your GCSE.
So it's important that you prepare well and feel good on the day.

Plan your revision in the weeks leading up to the exams.
Don't leave it too late!

The question 1's at the end of each chapter are a good way to revise the Summaries. These contain the essential notes you need.
Then try the past paper questions (coloured pages) on the chapter.
If you get stuck, ask a friend or your teacher the next day for help.

Just sitting there (especially in front of the TV!), reading your notes isn't good enough for most people. ***Active*** revision is better.
And don't try to revise for too long without a break.
Do 25 minutes, then promise yourself a 10 minute rest.
This works better than trying to revise non-stop.

So you've finished your revision (it's too late to worry about that anyway!), and it's the day of the exam. What will you need?
Remember to bring:

- Two pens (in case one runs out).
- A pencil for drawing diagrams.
- An eraser and ruler.
- A watch for pacing yourself during the exam. (It might be tricky to see the clock in the exam room.)
- A calculator (with good batteries).

You will feel better if you know exactly what to expect.
So collect all the information about your exam papers.
You can use a table like the one shown below:

Date, time and room	Subject, paper number and tier	Length	Types of question: – structured? – single word answers? – longer answers? – essays?	Sections?	Details of choice (if any)	Approximate time per page
4th June 9.30 Hall	Science (Double Award) Paper 2 (Chemistry) Foundation Tier	$1\frac{1}{2}$ hours	Structured questions (with single-word answers and longer answers)	1	no choice	4–6 min.

In the exam

Make sure you read the front of the exam paper carefully.
Look at the exam cover opposite:
How is your exam paper different to this one?

Here are some hints on answering questions in the exam:

Answering 'structured' questions:

- Read the information at the start of each question carefully. Make sure you understand what the question is about, and what you are expected to do.
- Pace yourself with a watch so you don't run out of time. If you have spare time at the end, use it wisely.
- ***How much detail do you need to give?***
  The question gives you clues:
  - Give short answers to questions which start: '**State** . . .' or '**List** . . .' or '**Name** . . .'.
  - Give longer answers if you are asked to '**Explain** . . .' or '**Describe** . . .' or asked '**Why does** . . ?'.
- Don't explain something just because you know how to! You only earn marks for doing exactly what the question asks.
- Look for the marks awarded for each part of the question. It is usually given in brackets, e.g. [2]. This tells you how many points the examiner is looking for in your answer.
- The number of lines of space is also a guide to how much you are expected to write.
- Always show the steps in your working out of calculations. This way, you can gain marks for the way you tackle the problem, even if your final answer is wrong.
- Try to write something for every part of each question.
- Follow the instructions given in the question. If it asks for one answer, give only one answer. Sometimes you are given a list of alternatives to choose from. If you include more answers than asked for, any wrong answers will cancel out your right ones!

National Examining Board

SCIENCE:

CHEMISTRY
Foundation Tier

4th June 9.30 a.m.

Time: 1 hour 30 minutes

Answer **ALL** the questions.

In calculations, show clearly how you work out your answer.

Calculators may be used.

Mark allocations are shown in the right-hand margin.

In what ways is your examination paper different from this?

Glossary

Acid When dissolved in water, a solution of an acid has a pH number less than 7. It is the chemical opposite of an alkali.

Alkali A base that will dissolve in water. Its solution has a pH number greater than 7. It is the chemical opposite of an acid.

Atom The smallest particle of an element that can still be called the element. All atoms contain protons, electrons and neutrons.

Base The oxide, hydroxide or carbonate of a metal. If a base dissolves in water it is called an *alkali*.

Catalyst A substance that speeds up a reaction without being chemically changed itself at the end of the reaction.

Chemical change A change (reaction) in which one or more new substances are made, eg. wood burning.

Combustion The reaction in which a substance burns in oxygen (or air).

Composition The type and amount of each element in a compound.

Compound A substance made when two or more types of atom are chemically bonded together.

Displacement When one element takes the place of another in a compound. For example,

magnesium + copper sulphate →
magnesium sulphate + copper

Electron A tiny particle with a negative charge which orbits the nucleus of an atom in shells or energy levels.

Element A substance that is made of only one type of atom.

Endothermic A reaction that *takes in* heat energy from the surroundings.

Enzymes Biological catalysts which speed up reactions in plants and animals.

Equation A shorthand way of showing the changes that take place in a chemical reaction.

eg. hydrogen + oxygen → water (a word equation)

$2H_2 + O_2 \rightarrow 2H_2O$ (a balanced equation)

Erosion The wearing away of rocks by other bits of rock moving over them.

Exothermic A reaction that *gives out* heat energy to the surroundings.

Fermentation The reaction when sugar (glucose) is turned into alcohol (ethanol) and carbon dioxide gas.

Fossil fuels A fuel made from plants or animals that died millions of years ago. eg. coal, oil, natural gas.

Global warming The build up of 'greenhouse' gases that is causing the average temperature of the Earth to rise.

Group All the elements in a column of the Periodic Table.

Igneous rock A rock formed when molten (melted) rock cools down and forms crystals.

Ion A charged particle (formed when atoms lose or gain electrons).

Lava Molten rock ejected from an erupting volcano.

Magma Molten rock below the Earth's surface.

Molecule A group of atoms chemically bonded together.

Neutral A solution or liquid which is neither acidic nor alkaline with a pH value of 7. (Also means 'carrying no electrical charge'.)

Neutralisation The chemical reaction of an acid with a base (alkali), in which they cancel each other out.

Neutron A dense, neutral particle found in the centre (nucleus) of an atom.

Non-renewable resources Energy sources that cannot be replaced once used up. eg. fossil fuels.

Nucleus The centre of an atom, containing protons and neutrons.

Oxidation A reaction in which oxygen is added to a substance.

Period All the elements across one row of the Periodic Table.

Periodic Table A table showing the chemical elements in order of their atomic numbers.

pH number A number which shows how strong an acid or alkali is.

Acids have pH values below 7; pH 7 is a neutral solution; alkalis have pH values above 7.

Physical change A change in which no new substance is made. eg. Ice melting.

Product A substance made in a chemical reaction.

Proton A dense, positively charged particle found inside the nucleus of an atom.

Reactant The substances that react together in a chemical reaction.

Reaction A chemical change which makes one or more new substances.

reactants → products

Reactivity series A list of metals in order of their reactivity. The most reactive metal is put at the top of the list.

Reduction A reaction in which oxygen is removed.

Sedimentary Rock A rock formed when materials settle out in water, building up in layers.

States of Matter Solid, liquid and gas are the three states of matter.

Thermal decomposition The breakdown of a substance by heat.

Index

Acknowledgements

I would like to thank the following people for their help and support in writing this book:

Sarah Coulson, Michael Cotter, Beth Hutchins, Derek McMonagle, Stewart Miller, Nick Paul, Claire Penfold, Mark Pinsent, Simon Read, Judy Ryan, and Susannah Wills.

Acknowledgement is made to the following Awarding Bodies for their permission to reprint questions from their examination papers:

AQA	Assessment and Qualifications Alliance
EDEXCEL	Edexcel Foundation
OCR	Oxford, Cambridge and RSA Examinations
WJEC	Welsh Joint Education Committee

Illustration acknowledgements

Advertising Archive: 6b; **Jon Arnold Images** (www.jonarnold.com): 17; **Art Directors & Trip:** 12, 14t, 20b, 21t and b, 22, 29t and c, 34, 37t, 52, 55b, 56c, 70b, 129, 147b, 167, 180, 196; **Biopol:** 92b; **J Allen Cash Photolibrary:** 140; **Martyn Chillmaid:** 21tb, 23, 55t, 58t, 78tcr, br and bcl, 86t, 90, 132, 156; **Bruce Coleman Collection**/K Burchett: 78bl, /CM Pampaloni: 56t; **Collections:** 26bc, 35t, 73l, /A Sieneking: 78bcr, 87; **Corbis UK Ltd:** 31r, 88b, 101b, 151, /Sygma: 10r, 73t; **James Davis Travel Photography:** 39; **Ecoscene:** 73r, 92t, 99, 153, /Papilio/Robert Pickett: 68; **Empics:** 31tl; **Mary Evans Picture Library:** 154t and b, 168; **Eye Ubiquitous:** 20t, 42c, 46b, 74, 136; **GeoScience Features Picture Library:** 26t, cl and bl, 101t, ca and cb, 102b, 104tl, tr and br; **Getty**/Food Pix: 14b, /Image Bank: 10t, 69, 72b, /Barros & Barros: 58b, /A van der Varen: 88t, /Stone: 31bl, 80, 89, 101, /M Brooke: 59r, /A Husmo: 76b, /C Keeler: 78cr, /H Staartjes: 82, /P Tweedie: 78tr, /K Wood: 78tcl; **Growbag**/Simon Roberts: 121l and r; **Hulton Getty:** 8b; **ICI:** 200b; **Image State:** 36, 46t, 59l, 139, 152, 186, 197l; **Impact Photos:** 35b, 201; **Frank Lane Picture Agency:** 78tl; **Jeff Moore** (jeff@jmal.co.uk): 8t, 13, 14c, 26br, 56b, 70t, 128, 149t and b; **PA News Photos:** 175; **Panos Pictures:** 42b; **Photos for Books** (info@photosforbooks.com): 18, 26cr, 37b, 42t, 50b, 51, 54, 72t, 141l, 197t, 202, 204t; **Pictor International:** 45, 96, 138; **Popperfoto:** 113b; **Railways-Milepost 92½:** 29 both; **Rex Features:** 86b, 141r, 184; **Lawrie Ryan:** 44; **Science Photo Library:** 6t, 9, 10c, 26c, 27, 30, 32, 50t, 67, 76t, 92c, 100, 103b, 108, 109, 113t, 114, 174, 200t, 204b; /Lawrence Livermore National Laboratory: 191, /W & D McIntyre: 197r; **Spectrum Colour Library:** 102; **Stock Boston**/Tom Wurl: 147t; **University of Pennsylvania Library**/Edgar Fahs Smith Collection: 190; **Tony Waltham Geophotos:** 103t, 115; **Wellcome Trust Medical Photographic Library:** 75.

While every effort has been made to contact copyright holders, the publishers apologise for any omissions, which they will be pleased to rectify at the earliest opportunity.

Picture research by Liz Moore (lm@appleonline.net)

Grateful thanks to Hannah Sherry, Peter Williams and the Chemistry Department at Benenden School for their help in producing some photographs.